Sea Water Corrosion of Stainless Steels — Mechanisms and Experiences

EARLIER VOLUMES IN THIS SERIES

European Federation of Corrosion Publications
NUMBER 19

A Working Party Report on

Sea Water Corrosion of Stainless Steels — Mechanisms and Experiences

Published for the European Federation of Corrosion by The Institute of Materials

THE INSTITUTE OF MATERIALS
1996

Book Number 663
Published in 1996 by The Institute of Materials
1 Carlton House Terrace, London SW1Y 5DB

British Library Cataloguing in Publication Data
Available on application

Library of Congress Cataloging in Publication Data
Available on application

ISBN 1-86125-018-5

Neither the EFC nor The Institute of Materials
is responsible for any views expressed
in this publication

Design and production by
PicA Publishing Services

Made and printed in Great Britain

Contents

European Federation of Corrosion Publications
Series Introduction

The EFC, incorporated in Belgium, was founded in 1955 with the purpose of promoting European co-operation in the fields of research into corrosion and corrosion prevention.

Membership is based upon participation by corrosion societies and committees in technical Working Parties. Member societies appoint delegates to Working Parties, whose membership is expanded by personal corresponding membership.

The activities of the Working Parties cover corrosion topics associated with inhibition, education, reinforcement in concrete, microbial effects, hot gases and combustion products, environment sensitive fracture, marine environments, surface science, physico–chemical methods of measurement, the nuclear industry, computer based information systems, corrosion in the oil and gas industry, and coatings. Working Parties on other topics are established as required.

The Working Parties function in various ways, e.g. by preparing reports, organising symposia, conducting intensive courses and producing instructional material, including films. The activities of the Working Parties are co-ordinated, through a Science and Technology Advisory Committee, by the Scientific Secretary.

The administration of the EFC is handled by three Secretariats: DECHEMA e. V. in Germany, the Société de Chimie Industrielle in France, and The Institute of Materials in the United Kingdom. These three Secretariats meet at the Board of Administrators of the EFC. There is an annual General Assembly at which delegates from all member societies meet to determine and approve EFC policy. News of EFC activities, forthcoming conferences, courses etc. is published in a range of accredited corrosion and certain other journals throughout Europe. More detailed descriptions of activities are given in a Newsletter prepared by the Scientific Secretary.

The output of the EFC takes various forms. Papers on particular topics, for example, reviews or results of experimental work, may be published in scientific and technical journals in one or more countries in Europe. Conference proceedings are often published by the organisation responsible for the conference.

In 1987 the, then, Institute of Metals was appointed as the official EFC publisher. Although the arrangement is non-exclusive and other routes for publication are still available, it is expected that the Working Parties of the EFC will use The Institute of Materials for publication of reports, proceedings etc. wherever possible.

The name of The Institute of Metals was changed to The Institute of Materials with effect from 1 January 1992.

A. D. Mercer
EFC Scientific Secretary,
The Institute of Materials, London, UK

EFC Secretariats are located at:

Dr J A Catterall
European Federation of Corrosion, The Institute of Materials, 1 Carlton House Terrace, London, SW1Y 5DB, UK

Mr R Mas
Fédération Européene de la Corrosion, Société de Chimie Industrielle, 28 rue Saint-Dominique, F-75007 Paris, FRANCE

Professor Dr G Kreysa
Europäische Föderation Korrosion, DECHEMA e. V., Theodor-Heuss-Allee 25, D-60486, Frankfurt, GERMANY

Preface

It is our privilege, on behalf of SINTEF and the EFC to introduce these proceeding of the European Workshop in Trondheim on "Sea Water Corrosion of Stainless Steel — Mechanisms and Experiences".

This workshop was organised by SINTEF Corrosion and Surface Technology Centre in co-operation with the EFC (European Federation of Corrosion) working parties on Microbial Corrosion and Marine Corrosion. It was also a closing seminar for the European Commission MAST (Marine Science and Technology) project "Marine Biofilms on Stainless Steels" and hence a symbol of European scientific and industrial co-operation.

Norway has among the longest coastlines in European and that is one of the reasons why marine activities always have been an important part of the Norwegian economy. Earlier, whale and seal hunting, fishing and shipping were most important; today it is offshore oil and gas production. It is therefore appropriate for this Workshop to be organised at a Norwegian venue.

Making a living from the marine environment is still a challenge to men and women. The future oil and gas fields in the North Sea are in greater water depths and the wells to be drilled are deeper; the shipping industry is looking for new and faster boats and the fishing industry is moving from harvesting to farming. Common to all these activities is the need for materials which are able to resist the corrosive environment which sea water and the marine atmosphere represent.

The introduction of the new highly alloyed stainless steels in the early 1980s was considered by many to offer a solution to the corrosion problems in sea water handling systems. The Norwegian oil companies were among the first to put these steels into use and, as with all new developments, the experiences have been both promising and disappointing — it takes time to understand the possibilities and limits that a new material presents.

The workshop has demonstrated that selection of stainless steels for sea water applications is still a challenge and an area of learning. It is particularly fascinating to know that the rich marine biolife also affects the corrosion properties of these materials. This phenomenon was reported in the literature by Italian (1976) and Norwegian (1983) scientists from ICMM and SINTEF, respectively. The first observation of quite noble potentials on stainless steels in sea water was, however, recorded as early as 1940, but no attempt was made at that time to explain the observation. The mechanism behind the ennoblement is still not fully understood and we hope that this meeting — at which scientists from ICMM and other European Sea Water Laboratories and research institutions were present — will have increased our understanding of the effect of biofilms on stainless steels.

The first day of this workshop focused on the mechanisms behind the microbially influenced corrosion of stainless steels, with the presentation of results from two joint European projects. The practical consequences of the biofilms with respect to corrosion were also addressed by a number of speakers. The second day concentrated entirely on industrial applications of stainless steels and the experiences of end-users and steel manufacturers. Where and how to use stainless steels in sea water systems and new solutions to combat corrosion of stainless steels in sea water systems were also discussed.

The organisers were particularly pleased to have important end-users and steel manufacturers present at this meeting to participate in the discussions, and we can look forward to further valuable exchange of information between manufacturers, test laboratories and end-users. If European industry to be competitive in the future, the time from concept to industrial application has to be reduced. To achieve this aim the communication between scientists, engineers and the industry has to be improved. We all have a lot to learn from each other, and we hope that this meeting and these proceedings will be an example of a fruitful exchange of knowledge and experience.

Unni Steinsmo   **Bard Espelid**   **Dominique Thierry**
SINTEF   Chairman *EFC Working Party*   Chairman *EFC Working Party*
                     on Marine Corrosion   *on Microbial Corrosion*

Trond Ronge
SINTEF

1

Comparison of Sea Water Corrosivity in Europe

K. P. FISCHER*, E. RISLUND†, O. STEENSLAND**, U.STEINSMO††
and B. WALLÉN§

*Marintek, Sandefjord, Norway
†Force Institutes, Denmark
**Det Norske Veritas, Høvik, Norway
††SINTEF, Norway
§Avesta Sheffield, Sweden

ABSTRACT

A Round Robin Test has been performed to map differences in sea water corrosivity towards stainless steel among European laboratories. The sea water varied from brackish to full strength ocean water. The test comprised monitoring of free corrosion potential and the reduction current of UNS S31254, crevice corrosion testing of three different stainless steel qualities, comparison of critical pitting temperature of UNS S31600 and monitoring of sea water parameters. Potential ennoblement and increase in the reduction current were observed at all the test stations involved during both seasons. The rate of the anodic reaction in the stable pitting process was practically independent of the sea water composition. The tendency towards initiation of crevice corrosion was found to differ significantly among the test stations involved, which is difficult to explain based on data available on chemical, physical properties and biological effects.

1. Introduction

Sea water is used for a range of industrial activities such as transport, exploration of natural resources, power production, water supply, etc. Sea water is a highly corrosive environment and is still a challenge to corrosion experts. A major basis for material selection is the results from standardised and specially designed corrosion tests. In some tests, natural sea water is used as electrolyte. The corrosivity of natural sea water is related not only to its chemical and physical properties, but also to its biology [1–14], and the characteristics of sea water will vary among geographical locations. In 1992 the EFC Working Party on Marine Corrosion initiated a Round Robin Test to map differences in sea water corrosivity towards stainless steels among European laboratories. Eleven test stations participated in the investigation, representing eight different countries. This paper summarises the results from the test.

2. Experimental

The Round Robin Test comprised the following activities:

1. Monitoring the free corrosion potential and the reduction current of stainless steel UNS S31254 exposed in sea water as a function of time.
2. Crevice corrosion testing of three different stainless steel qualities: UNS S31600, UNS S31254 and UNS N08904.
3. Comparison of critical pitting temperature of UNS S31600 in sea water obtained from different test stations.
4. Monitoring of sea water parameters during exposure periods, i.e. temperature, pH, oxygen and chloride concentration, conductivity and redox potential.

To study the importance of seasonal variations, the testing was carried out in two consecutive periods, each of half a year duration.

The sea water laboratories involved are given in Table 1 together with data on sea water quality. The critical pitting test was performed by Avesta Sheffield using the Avesta Cell. The potential independent critical pitting temperature, defined as the temperature when the breakthrough potential at $100 \ \mu A \ cm^{-2}$ drops from the transpassive to the pitting region, was determined using a scan rate of $20 \ mV \ min^{-1}$ [15,16]. Further experimental details are given in an earlier presentation and will also be available in a separate report [17].

Table 1. EFC Marine Working Party — Round Robin Test on Sea water Corrosivity — Participating Laboratories[1]

Test station (country)[2]	Sal. ‰ [3]	Test I[4]		Test II [5]	
		Initial temp., °C	Average temp., °C	Initial temp., °C	Maximum temp., °C
Zagreb Univ., Dubrovnik (Croatia)	37.6	25	18	17	26
SINTEF, Trondheim (NO)	34.3	10	11	6	12
Marintek, Sandefjord (NO)	34	9	6	10	17
IFREMER, Brest (F)	34	15	≈ 12	11	15
DNV, Bergen (NO)	32.9	10	9	8	11
CEA, Cherbourg (F)	32.5	19	12	12	24
SCI, Stockholm (S)	32.1		5[6]	4	17
Naval College, Den Helder (NL)	27.6	19	13	7	23
FORCE Inst., Copenhagen (D)	< 20			13	24
CTO, Gdansk (P)	6.7	17	11	6	26
HelsinkiUniv. (F)	4.8	16	6	1	17

[1] Average pH: 7.5–8.2. Average oxygen conc.: 5.8–8.8.

[2] F = France, NO = Norway, S = Sweden, P = Poland, F = Finland, D = Denmark, NL = Netherlands.

[3] Salinity calculated from data on chloride content using the equation: S ‰ = 0.03 + 1.805*[Cl-].

[4] Test start week 26–47.

[5] Test start week 9–24.

[6] Final temperature.

3. Results and Discussion
3.1. The Free Corrosion Potential of UNS S31254

The free corrosion potential of UNS S31254 in natural sea water has been measured as a function of time and, as illustrated in Fig. 1, the data are quite similar in spite of the differences in sea water quality (Table 1).

The change in potential vs time is characterised by an initial delay time of 40–120 h prior to a sharp potential increase to values about 100–380 mV SCE for a period of 10–20 d followed by a slow increase to a maximum value about 352–520 mV SCE. (Data from Test II for all laboratories are given in Table 2.) The potential levels measured are generally in accordance with data reported for passive stainless steel surfaces in natural sea water with microbial activity [1–14]. A few laboratories, however, have recorded higher potentials, i.e. > 450–500 mV SCE.

To study the effect of seasonal variations a winter test (Test I) and a summer test (Test II) were performed. Comparing the data from the two test periods, there seems to be a general tendency indicating that the potential increase rate is lower in the summer test (which started in early spring with low sea water temperature) than in the winter test (started in mid-summer with high sea water temperature), as illustrated in Fig. 2. The laboratories at DNV and Marintek showed the only exception, at which the potential increase rate was the same for the two test periods investigated.

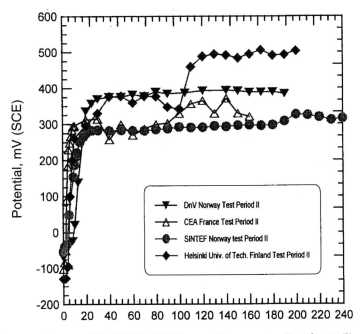

Fig. 1 Free corrosion potential of UNS S31254 exposed to sea water. Data for sterilised water is shown for comparison [2, 9].

Table 2. *Free corrosion potential and the reduction current of UNS S31254 at different test laboratories (Test II)*

Laboratory	Potential increase started after (h)	Potential increase rate (mV h⁻¹)	Final potential (mV SCE)	Current increase[1] after (h)	Max.[1] current (µA cm⁻²)
Zagreb U.	42	1.4	350	840	−0.25
SINTEF	55	1.7	348	330	−10.3
Marintek	85	3.28	520	240	−8.1
IFREMER	48	3.1	410	400	−10
DNV	123	2.17	380	470	−9.8
CEA	48	6	350	96	−6
SCI	88	0.9	443	460	−3.0
Naval College	120	1.74	500	360	−9.2
FORCE	37	3.2	500–550	188	−1
CTO		0.8	352	480	−2.0
Helsinki Univ.	100	3.6	500	296	−1.0

[1] = Absolute value

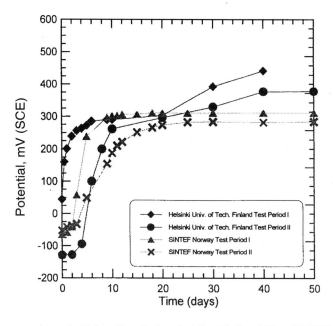

Fig. 2 *Free corrosion potential vs time during Test Periods I and II at SINTEF; Norway and Helsinki University.*

3.2. The Reduction Current of UNS S31254

The cathodic current density of UNS S31254 at the polarisation potential 0 mV SCE was also monitored as a function of elapsed time. The overall tendency is a delay time followed by a sharp current increase to a maximum absolute value and then a decline in current. The delay time in Test II varied from 96 to 840 h of exposure, and is generally longer than the delay time observed for the potential.

Hence, at all test stations there is an increase in the absolute value of the current which is not observed in sterilised/artificial sea water [1]. The reduction current in sea water without active biofilm at the potential 0 mV SCE is typically of the order of 0.001 μA cm^{-2}. The maximum absolute value of the reduction current measured in natural sea water varied between 0.14–15.4 μA cm^{-2} in Test I and 0.25–10.3 μA cm^{-2} in Test II. The scatter in test data obtained at the different European Sea water Test Laboratories is illustrated in Fig. 3. (Data from Test II are given in Table 2.) The lowest currents (i.e. max. absolute value < 5 μA cm^{-2}) are during test period II, measured at Helsinki University, SCI, Force Institute, CTO and Zagreb University. Maximum absolute values > 5 μA cm^{-2} were measured at all the three test stations in Norway, at CEA, IFREMER and Naval College. The data from Test I are similar, with the exception of results at CEA and IFREMER, which were < 5 μA cm^{-2} in this test.

The highest currents were measured at laboratories with high salinity water, and the current densities were low at both the laboratories with brackish water. Low currents, however, were also recorded at SCI, Zagreb University and Force Institutes.

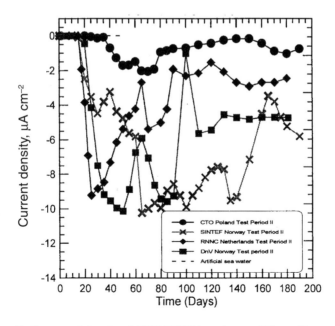

Fig. 3 *The cathodic current density of UNS S31254 vs time at different European sea water laboratories. The polarisation potential is 0 mV SCE.*

Table 3. Initiation of crevice corrosion at the different test laboratories

Alloy	UNS S31600		UNS N08904		UNS S31254	
Test	Test I	Test II	Test I	Test II	Test I	Test II
Zagreb Univ.	2/4	4/4	0/4	3/4	0/4	0/4
SINTEF	4/4	3/4	1/4	3/4	0/4	0/4
Marintek	1/4	2/4[2]	0/4	0/4	0/4	0/4
IFREMER	0/4	0/4	1/4	0/4	0/4	0/4
DNV	0/4	0/4[1]	0/4[2]	0/4[1]	0/4	0/4
CEA	3/4	4/4	1/4[3]	4/4	0/4	0/4
SCI	1/4	0/4[1]	1/4	0/4	0/4	0/4
Naval College	4/4	3/4	0/4	3/4	0/4	0/4
FORCE		4/4		3/4		0/4
CTO	2/4[1]	3/4	0/4[1]	2/4	0/4	0/4
Helsinki Univ.	0/4	0/4	0/4	0/4	0/4	0/4
Total	17/40[1]	22/44[4]	4/40[6]	19/44[1]	0/40	0/44

[1,2,3,4,6] In addition, 1/2/3/4/6 specimens respectively initiated at cable connection. Due to the potential drop, initiation at cable connection might prevent corrosion occurrence at the "less severe crevice" created by the plastics crevice formerly used.

3.3. Comparison of Critical Pitting Temperature of UNS S31600

It is also possible that the variation in chemical composition of the sea water may influence its corrosivity. Important parameters here are the chloride concentration and the pH, which both may affect the susceptibility to localised corrosion, i.e. crevice or pitting corrosion. To investigate if the sea water composition has an effect on corrosion initiation, the pitting resistance of one and the same stainless steel was electrochemically determined in sea water of different origin. The sea water sampled at the different laboratories ranged from brackish to full strength ocean water (as can be seen from Table 1). The salinity corresponds to chloride concentrations of approximately 3–20 000 ppm. In spite of the differences in conductivity, the CPT determined in the different waters are very much the same, varying from 28°C to 33–34°C in both the winter and summer test. The variations observed are almost within the reproducibility of the test method.

3.4. Crevice Corrosion Testing

The corrosivity of sea water towards stainless steels is greatly influenced by the cathodic reaction of the corrosion process. It has been shown that the rate of this reaction is very much controlled by the existence of a biofilm on the steel surface. The biofilm catalyses the oxygen reduction, causing the free corrosion potential and the reduction current to increase, when polarising the steel specimen in a cathodic direction. In this way the biofilm increases both the risk of localised corrosion initiation and the rate of propagation. Parameters to be considered when evaluating the corrosivity of sea water towards stainless steels should therefore include (i) temperature, (ii) potential level, (iii) potential increase rate, and (iv) cathodic reduction rate.

Looking at the data from the crevice corrosion testing, there seems to be no clear connection between these parameters and the tendency to crevice corrosion. The sea water at the IFREMER test station and SINTEF are comparable with respect to potential level and reduction current during Test II. The temperature and potential increase rate are higher at IFREMER. In spite of this, crevice corrosion was registered on 3 out of 4 test specimens of UNS S31600 at SINTEF and on none of the test specimens at IFREMER. Crevice corrosion initiation is a stochastic process, and further statistical analyses have to be performed in order for the probability of the two cases mentioned to occur (assuming identical test specimens). In other cases the variation among the laboratories may be attributed to differences in sea water temperature. This can explain the results from SCI, Helsinki University and DNV.

There would appear to be a significant variation among the test laboratories involved with respect to initiation of crevice corrosion — which is difficult to completely explain from the data available on chemistry, biological effects and temperature. Using the extent of crevice corrosion on UNS S31600 as the criteria, the laboratories can be divided in three corrosivity groups as summarised in Table 4.

4. Conclusions

Potential ennoblement of stainless steel occurred at all the test stations involved in the Round Robin test during both the seasons investigated.

An increase in the reduction current at 0 mV SCE was observed at all test stations. The increase in the magnitude of the current was delayed compared to the potential increase. The magnitude of the current was found to differ among the laboratories involved.

Electrochemical determination of critical pitting temperature and pitting potential of UNS S31600 in different sea water shows that the rate of the anodic reaction in the stable pitting corrosion process is practically independent of the composition of sea water (ranging from brackish to full strength ocean water).

The tendency towards initiation of crevice corrosion was found to differ significantly among the test stations involved; for UNS S31600, the variation ranged

Table 4. Classification of the test stations involved; based on intiation of crevice corrosion on UNS S31600

	Definition	Laboratories
I	Max. one of eight specimens attacked	Helsinki Univ., DNV, SCI, IFREMER
II	≤ 50% of specimens attacked	Marintek
II	> 50% of specimens attacked	FORCE, CEA, Naval College, Zagreb Univ., SINTEF, CTO

from 0 of 8 test specimens at three test stations to a maximum of 7 of 8 at other stations. The differences observed are difficult to explain, based on data available on chemical, physical properties and biological effects.

References

1. R. Holthe, The cathodic and anodic properties of stainless steels in sea water, Dr. Ing. thesis, Univ. of Trondheim, Norway, 1988.

2. V. Scotto, R. DiCintio and G. Marcenaro, *Corros. Sci.*, 1985, **25**, 185.

3. A. Mollica and A. Trevis, *Proc. 4th Int. Congr. on Marine Corrosion and Fouling*, Juan-les Pins, 1976.

4. E. D. Mor, U. Scotto and A. Mollica, *Werkstoffe und Korros.*, 1980, **31**, 281.

5. A. Mollica, A. Trevis, E. Traverso, G. Venture, V. Scotto, G. Alabiso, G. Marcenaro, U. Montini, G. De Carolis and R. Dellepiane, *Proc. 6th Int. Congr. on Marine Corrosion and Fouling*, Athens, 1984, p.269.

6. F. L. La Que and G. L. Cox, *Proc. ASTM* , 1940, 40.

7. R. Johnsen, E. Bardal and J. M. Drugli, *Proc. 9th Scand. Corros. Congr.*, Copenhagen, September 1983.

8. R. Johnsen and E. Bardal, *Corrosion*, 1985,**41**, 296.

9. R. Johnsen and E. Bardal, *Corrosion '86*, Paper No. 227, NACE, Houston, Tx, USA, 1986.

10. E. Bardal and R. Johnsen, *Proc. UK Corrosion '86*, Brimingham, 1986, p.287.

11. S. Valen, E. Bardal, T. Rogne and J. M. Drugli, 11th Scand. Corros. Congr., Stavanger, June 1989.

12. P. O. Gallagher, R. E. Malpas and E. B. Shone, *Brit. Corros. J.*, 1988, **23**, 4.

13. J. P. Audouard, A. Desestret, L. Lemoine and Y. Morizur, *UK Corrosion '84*, Wembley Conf. Centre, London, UK, 1984.

14. S. C. Dexter and G. Y. Gao, *Corrosion*, 1988, **44**, 717.

15. R. Qvarfort, *Corros. Sci.*, 1988, **28**, 2, 135.

16. P. E. Arnvig and R. M. Davison, *Proc. 12th ICC Conf.*, Houston, Tx, 19–25 Sept., 1993, 1477.

17. WP on Marine Corrosion, Comparison of sea water corrosivity, EFC report to be published.

2

Effect of Marine Biofilms on Stainless Steels: Results from a European Exposure Program

J. P. AUDOUARD, C. COMPÈRE*, N J. E. DOWLING, D. FÉRON†,
D. FESTY*, A. MOLLICA**, T. ROGNE††, V. SCOTTO**, U. STEINSMO††,
C. TAXEN§ and D. THIERRY§

IRSID, BP 01, 42490 Fraisses, France
* IFREMER, BP 70, 29280 Plouzané, France
† CEA-LETC. Etab. de la Hague, 50444 Beaumont-Hague, France
** CNR/ICMM, Torre di Francia, Via de Marini, 6-lVp.,16149 Genova, Italy
††SINTEF- Corrosion Center, Rich. Birkelandsv. 3a, N-7034, Trondheim, Norway
§ Swedish Corrosion Institute, RoslagsvŠgen 101, Hus 25, 10405 Stockholm, Sweden

ABSTRACT

During the last decade, the use of stainless steel for applications in marine environments has increased extensively. In particular, stainless steels are widely used in power plants and off-shore structures. However, as a result of biofilm adhesion in natural marine environments, the open-circuit potential of stainless steel is displaced to a value that may be close to the breakdown potentials for several of these materials. Consequently, in 1992, the Marine Science and Technology (MAST) Directorate in Brussels, Belgium,initiated a collaborative program between seven research institutes in Europe to investigate the effects and properties of marine biofilms which develop on high alloyed stainless steels in the seas immediately surrounding the western European coastline. The results show that biofilm settlements on the stainless steel surface cause an increase in the open-circuit potential, independent of the geographical location of the marine station. Furthermore, the open-circuit potentials reach a stationary value with time, independent of the stainless steel quality, the geographical location and the hydrological parameters of the marine stations. The biofilms which formed on stainless steel at different open-circuit potentials and at different locations have been collected and their weight and composition (extracellular polymer, proteins and intracellular carbohydrates) determined.

1. Introduction

Stainless steels have been used as construction materials in marine environments for many decades. In particular, stainless steels are widely used by the oil industry for offshore equipment. The main factor limiting the use of lower grade stainless steels (such as AISI 304 and AISI 316) in sea water is their susceptibility to localised corrosion in the form of pitting and/or crevice corrosion. It has been shown by several groups that the open circuit potential (i.e. the corrosion potential) is shifted more towards the noble direction for stainless steel exposed to natural sea water than for stainless steel exposed to artificial sea water [1–4]. This effect, also referred to in the literature as "potential ennoblement", has been observed for different sea waters having different salinities. However, conflicting results have been reported in some cases, showing the same evolution of the corrosion potential in natural sea water as in artificial sea water [5, 6]. The reasons for this discrepancy are unclear.

Two different mechanisms have been proposed for the shift in corrosion potential: an increase in the rate of the oxygen reduction reaction, and a shift in the reversible potential for oxygen reduction. It is also possible that these two act together (see, for instance, a review of the mechanisms [7]). However, the mechanism(s) responsible for the potential ennoblement is not known, although in all cases it has related to the presence of a functional biofilm on the stainless steel surface. The effect of the potential ennoblement on the corrosion resistance of stainless steels in sea water may be summarised as follows:

(1) An increase the susceptibility to the initiation of crevice corrosion by increasing the potential to a value near or above the breakdown potentials of several of these alloys;

(2) An increase in the rate of propagation of crevice corrosion by the catalysis of oxygen reduction on the cathodic areas outside the crevice.

Results from seasonal exposure tests carried out in different European coastal waters are given here in order to establish whether the effects of marine biofilms on stainless steels varied with geographical location or hydrological properties of the sea water.

2. Experimental
2.1. Materials

Most of the data presented here come from two super-austenitic stainless steels (i.e. UR SB8 and 654 SMO) and a super-duplex stainless steel (i.e. SAF 2507). Some experiments were also conducted on other alloys. The composition of the different alloys is given in Table 1. The materials were tested both in tube and plate form.

2.2. Marine Stations and Exposure Conditions

The specimens were degreased by trichloroethylene vapour and pickled for 20 min in a solution of 20% HNO_3 and 2% HF prior to exposure to sea water. The specimens were exposed with or without crevice geometries. All experiments reported for 654 SMO were performed without crevice geometries. For all the other alloys, a crevice assembly was used with a gasket material of type POM and a torque of 6 Nm.

The materials were exposed in the coastal waters outside Trondheim (Norway), Kristineberg (Sweden), Cherbourg (France), Brest (France) and Genova (Italy). Figure 1 shows schematically the locations of the marine stations. Four different exposure

Table 1. Composition of the alloys in % (ND = not detectable)

Alloy	Element									
	C	S	P	Si	Mn	Ni	Cr	Mo ø	Cu	N_2
AISI 316L	0.017	ND	ND	ND	ND	12.6	17.2	2.6	ND	0.05
254 SMO	0.01	0.001	0.02	0.38	0.5	17.8	19.9	6	0.69	0.2
654 SMO	0.01	0.001	0.02	0.16	3.6	21.8	24.5	7.3	0.43	0.48
47 N	0.01	0.001	0.02	0.5	1.3	6.6	24.5	2.9	ND	0.18
52 N	0.01	0.0001	0.01	0.24	1.1	6.3	25	3.6	1.5	0.25
B 26	0.009	0.0004	0.01	0.22	0.8	24.7	20	6.3	0.8	0.19
SB8	0.01	0.0008	0.01	0.24	0.93	25	25	4.7	1.4	0.21
SAF 2205	0.15	0.001	0.017	0.5	0.7	5.5	22	3.2	ND	0.17
SAF 2507	0.14	0.001	0.015	0.27	0.4	6.9	24.9	3.8	ND	0.28
SAN 28	0.019	0.001	0.017	0.44	1.7	30.3	26.7	3.4	0.9	0.07

periods were chosen in winter, spring, summer and autumn, during the period January 1993–September 1994. The exposures were stopped when the corrosion potential reached its maximum value (between 10 and 45 d after the start of the exposure depending on the season and the geographical location of the marine station).

The corrosion potential and the temperature were recorded simultaneously every 4 or 6 h using data acquisition systems. Plate specimens were exposed to sea water in the dark in PVC tanks with a sea water renewal rate of about 2 L min⁻¹ for tank capacity of 100 L. Tube specimens were exposed in a once-through system using a flow rate of between 0.5 and 1 m s⁻¹.

2.3. Hydrological Parameters

The hydrological pararneters for all stations are given for the summer and winter periods in Tables 2 and 3, respectively. These parameters, and the open-circuit potential of platinum (redox potential), were measured several times during the course of the exposure.

2.4. Biofilm Collection and Analysis

The collection of biofilms and their analysis have been described in detail elsewhere [8]. The metal specimens (plates or tubes) were withdrawn from the exposure system and leant on one edge to draw off excess water, then weighed to determine the total wet biofilm mass. The biofilm was then removed from the metal surfaces by ultrasonic washing in 20 mL of a buffer solution. The resulting suspension was centrifuged at 5000 rpm in order to separate the exopolymeric component (soluble in the supernatant) from the organism and detritus (pellet). The two fractions were sent to

Fig. 1 *Schematic view of the geographical location of the marine stations.*

Table 2. *Hydrological conditions for all marine stations. Summer 1993 (ND = not defined)*

Marine Stations	Genova	Cherbourg	Brest	Kristineberg	Trondheim
Temperature, °C	24.6 ± 0.7	13.9 ± 0.2	16 ± 1	10 ± 2	11.0 ± 0.1
pH	8.2 ± 0.1	8.10 ± 0.05	8.2 ± 0.1	7.8	8.0 ± 0.1
Salinity, ‰	37.0 ± 0.2	30.0 ± 0.5	33 ± 1	32.7	34.0 ± 0.5
Chla, μg L^{-1}	0.34 ± 0.05	0.3	0.9 ± 0.4	<0.1	0.23 ± 0.05
O$_2$, ppm	5.2 ± 0.6	6.7 ± 0.3	8.5 ± 0.5	5.2	6.5 ± 0.5
Organic matter, mg L^{-1}	2.7 ± 0.4	1.4	0.7 ± 0.2	0.6	11 ± 1
Ash, mg L^{-1}	9.7 ± 0.5	ND	6.6 ± 0.5	10.5	ND

Table 3. *Hydrological conditions for all marine stations. Winter 1994. (ND = not defined)*

Marine Stations	Genova	Cherbourg	Brest	Kristineberg	Trondheim
Temperature, °C	13.3 ± 0.2	8.7 ± 0.3	8 ± 2	6.0 ± 0.5	10 ± 2
pH	8.05 ± 0.03	8.0 ± 0.1	8.1	7.8	8.10 ± 0.5
Salinity, ‰	37.1 ± 0.2	29.5 ± 0.5	33.7	30	33.0 ± 0.8
Chla, ug L^{-1}	0.11 ± 0.03	< 0.1	1.4 ± 1.1	0.2	0.12 ± 0.02
O_2, ppm	8.5 ± 0.6	11.0 ± 0.1	9.9	7.8	7.5 ± 0.5
Organic matter, mg L^{-1}	3.3 ± 0.4	1.0	1.6 ± 1.1	1	1.5 ± 0.5
Ash, mg L^{-1}	11.3 ± 2.4	ND	12 ± 6	0.3	ND

ICMM for biochemical analyses. The samples were examined for intracellular carbohydrate and extracellular polymeric (EPS) content using the antrone method [11]. The protein content was determined using the Lowry method [10].

3. Results
3.1. Evolution of the Open Circuit Potential for 654 SMO as a Function of the Season and Geographical Location

Figures 2 and 3 show the evolution of the open-circuit potential for 654 SMO tube specimens exposed at the different marine stations during the winter and summer periods, respectively. The results show corrosion potential ennoblement of stainless steel in natural sea water at all rnarine stations. The final value of the corrosion

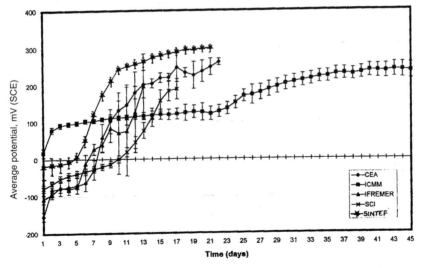

Fig. 2 *Evolution of the open-circuit potential exhibited by 654 SMO tubes at the five European marine stations during the winter exposure period. The results correspond to a mean value obtained on 5 to 20 specimens ± s.d.*

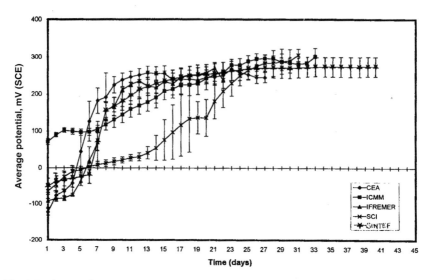

Fig. 3 *Evolution of the open-circuit potential exhibited by 654 SMO tubes at the five European marine stations dunng the summer exposure period. The results correspond to a mean value obtained on 5–20 specimens ± s.d.*

potential was similar in all cases, independent of the hydrological conditions and the season. However, the rate of increase and the initial corrosion potential show large differences between differen locations, and at different sea water temperatures. No correlation could be observed between the temperature of the sea water and the other properties displayed in Tables 2 and 3. Figures 4 and 5 show the evolution of the open-circuit potential for 654 SMO plate specimens exposed at the different marine stations during the winter and summer periods, respectively. Again, ennoblement of the corrosion potential was observed at all marine stations independent of the hydrological conditions and the season. However, the rate of increase and the initial corrosion potential show large differences between different locations, and at different sea water temperatures. Similar results have been obtained for both tube and plate specimens during exposures in spring and autumn seasons.

3.2. Evolution of the Corrosion Potential as a Function of the Composition of the Stainless Steel

Figure 6(a) shows the evolution of the corrosion potential for plate specimens of nine stainless steels mounted in crevice assemblies (see Table 1 for the composition) and exposed during the winter season at Genova. The typical s.d. is shown in Fig. 6(b). All stainless steels show similar ennoblement of corrosion potential, independent of their compositions (nickel, chromium and molybdenum content) or structure (austenitic or duplex stainless steels). It should be pointed out that the results in Fig. 6(a) refer only the specimens on which crevice corrosion was not observed (i.e. crevice

Average Potential vs Time - Winter 1994 (Plates SMO654)

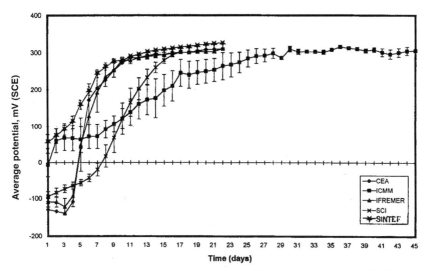

Fig. 4 *Evolution of the open-circuit potential exhibited by 654 SMO plates at the five European stations during the winter exposure period. The results correspond to a mean value obtained on 5 to 20 specimens ± s.d.*

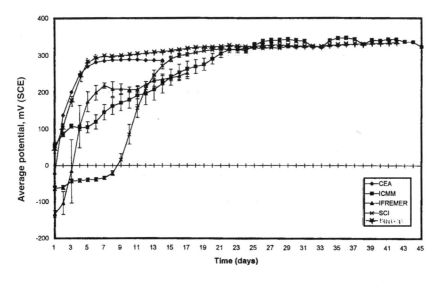

Fig. 5 *Evolution of the open-circuit potential exhibited by 654 SMO plates at the five European stations during the summer exposure period. The results correspond to a mean value obtained on 5 to 20 specimens ± s.d.*

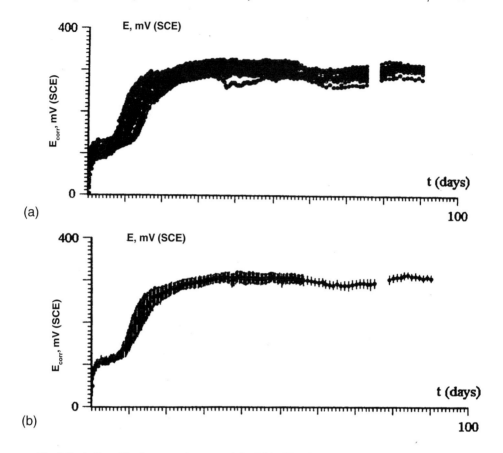

Fig. 6 Evolution of the free corrosion potential exhibited by plate specimens all the alloys described in Table 1 at the Genova marine station during the winter exposure period. (a) All alloys, (b) typical s.d. The results correspond to a mean value obtained on 5 specimens ± s.d.

corrosion was observed on some specimens of AISI 316L). The results of the crevice corrosion test is outside the scope of this paper and will be reported elsewhere. Similar results were obtained at all marine stations. (See comment on crevice corrosion in Discussion.)

3.3. Wet Weight of the Biofilm

Figure 7 shows the variation of the final corrosion potential for 654 SMO tube specimens exposed at marine stations as a function of the wet weight of the biofilm for a given exposure time. No correlation could be observed. Indeed, the mass of the biofilm varied by a factor of 100 between the minimum (at Cherbourg and Kristineberg) and the maximum at Genova. Thus, ennoblement of the corrosion potential is not directly related to the biomass but rather to the activity of the biofilm.

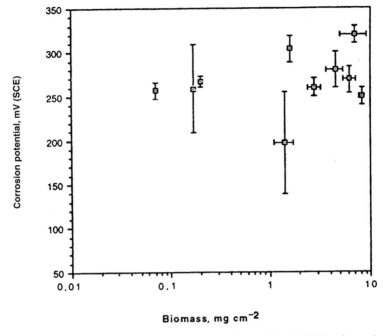

Fig. 7 *Relation between the final corrosion potential and the biomass for 654 SMO tube specimens. All marine stations. Winter and summer season exposures. The results correspond to a mean value obtained from 5 specimens ± s.d.*

3.4. Biofilm Analysis

The biofilms formed on 654 SMO at the different marine stations were analysed with respect to the intracellular carbohydrate, protein and extracellular polymeric substances (EPS) content. Figure 8 shows the variation of the EPS content of the biofilm as a function of the corrosion potential for tube and plate specimens of 654 SMO, SAF 2507 and SB8. These results include the data from all stations for exposure at all seasons, and show that a minimum amount of EPS is necessary (about 100 ng cm^{-2}) in order to obtain an increase of corrosion potential. Furthermore, the amount of EPS found in biofilms at the different marine stations seems to be in the same range, despite their very different biomasses. There also appears to be a relationship between the amount of EPS and the corrosion potential as shown in Fig. 9 which refers only to data obtained for biofilms formed at Genova. However, when considering all data (see Fig. 8), the relationship still exists for potentials above +150 mV SCE . Similar relationships were not observed for the intracellular carbohydrate and protein content of the biofilms. In fact, the amounts of thes substance varied greatly between the stations. Table 4 shows an example of the variation in these parameters between the different marine stations for the summer and winter exposure of tube specimens of 654 SMO at all stations. Table 5 shows the variation of the protein and intracellular carbohydrate together with the EPS content for the summer exposure of tube and plate specimens of 654 SMO at Genova.

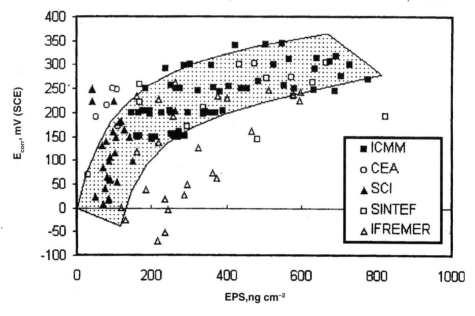

Fig. 8 *Variation of the EPS content of the biofilm formed on tube and plate specimens of 654 SMO SAF 2507 and SB8 as a function of the corrosion potential. All stations. All exposure periods. The dashed area corresponds to 80% of all experimental data.*

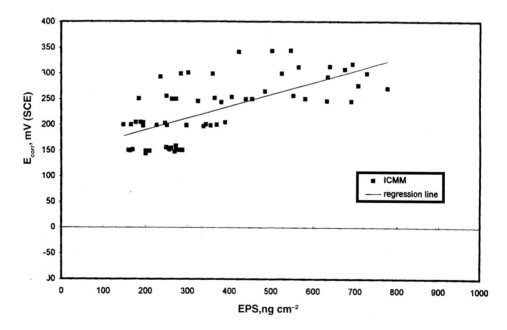

Fig. 9 *Same as Fig. 8, but for the marine station at Genova only. Correlation factor 0.7.*

Table 4. Hydrological conditions for all marine stations. Winter 1994. (ND = not defined)

Marine Stations	Genova	Cherbourg	Brest	Kristineberg	Trondheim
Temperature, °C	13.3 ± 0.2	8.7 ± 0.3	8 ± 2	6.0 ± 0.5	10 ± 2
pH	8.05 ± 0.03	8.0 ± 0.1	8.1	7.8	8.10 ± 0.5
Salinity, ‰	37.1 ± 0.2	29.5 ± 0.5	33.7	30	33.0 ± 0.8
Chla, μg L^{-1}	0.11 ± 0.03	< 0.1	1.4 ± 1.1	0.2	0.12 ± 0.02
O$_2$, ppm	8.5 ± 0.6	11.0 ± 0.1	9.9	7.8	7.5 ± 0.5
Organic matter, mg L^{-1}	3.3 ± 0.4	1.0	1.6 ± 1.1	1	1.5 ± 0.5
Ash, mg L^{-1}	11.3 ± 2.4	ND	12 ± 6	0.3	ND

Table 5. Protein intracellular carbohydrate and EPS content of biofilms formed on plate or tube of 654 SMO during the summer exposure at Genova

Specimen	Corrosion potential, mV ECS	Protein, μg cm^{-2}	Intracellular carbohydrate, ng cm^{-2}	EPS ng cm^{-2}
Tube	149 ± 4	2.6 ± 0.6	434 ± 44	210 ± 4
Tube	205 ± 6	7 ± 1	1260 ± 200	390 ± 27
Tube	277 ± 3	20 ± 4	4000 ± 400	707 ± 43
Tube	300 ± 3	17 ± 2	3400 ± 300	728 ± 80
Plate	155 ± 3	0.5 ± 0.1	227 ± 37	262 ± 37
Plate	199 ± 3	1.1 ± 0.3	174 ± 41	252 ± 23
Plate	247 ± 5	1.9 ± 0.4	310 ± 57	632 ± 110
Plate	313 ± 2	3.2 ± 0.8	364 ± 100	639 ± 130
Plate	343 ± 1	2.2 ± 0.3	426 ± 100	503 ± 70

4. Discussion
4.1. Potential Ennoblement

Potential ennoblement of stainless steel was observed in all cases, independent of the geographical location of the exposure site and the season of the exposure. This concurs with many earlier observations; passive stainless steels rapidly develop a high (and relatively stable) open-circuit potential in natural sea water [1–4]. Furthermore, it has been demonstrated that the ennoblement of the corrosion potential occurs independently of the nickel, chromium or molybdenum content of the alloy and its microstructure in accord with previous results, indicating that the biofilm accumulation with time was similar for stainless steels tubes of 6S4 SMO, SAF 2507 and SB8 [8].

Moreover, the corrosion potential reaches a steady plateau value at about 300–350 mV/SCE after a given time (see Figs 2–5), which seems to be independent of the geographical position and the hydrological characteristics of the marine stations.

However, the plateau value depends on the flow conditions of water, which was systematically lower for tube materials than for plate materials of 654 SMO for a given exposure period. These results may be explained by a different growth rate of the biofilm caused by the flow conditions. Plate materials experienced almost stagnant conditions, while there was a strong convective flow through the tubes.

These results have implications for the initiation and propagation of crevice corrosion, since the final value of the corrosion potential is rather similar at all stations for a given flow condition. The probability initiation of crevice corrosion for a given stainless steel alloy will be quite identical at all marine stations However, the rate of propagation of crevice corrosion will also depend on the temperature of the sea — which influences the kinetics of the oxygen reduction reaction outside the crevice area and, consequently, the anodic current density of the crevice area. It is, however, obvious that all the alloys studied in this project will not suffer crevice corrosion in natural sea water. The likelihood is that only AISI 316L will initiate corrosion under the actual testing conditions. Preliminary results on a crevice exposure test performed at all marine stations with the alloys described in Table 1 confirm this conclusion.

Although the final corrosion potential was rather similar for all stations, other properties varied widely. The initial corrosion potential, the time lag before the potential increase and the rate of increase of corrosion potential all showed a wide range of values. We cannot explain the observed differences, but can speculate on possible causes such as:

(1) a differing amount and composition of organic matter and/or EPS in the sea water may influence the corrosion potential of stainless steel; and

(2) a differing growth rate for the biofilms at the different stations can cause different rates of change in the corrosion potential.

This last point may be better understood when considering the differences in sea water introduction for the stations. At Cherbourg, Brest and Genova the sea water is pumped at the surface of the sea, while at Kristineberg and Trondheim the water is pumped at 60 and 40 m depth, respectively. While the temperature differential due to season at these depths is not large (see Tables 2 and 3), microflora will certainly be very different compared to the other stations.

4.2. Relation between the Potential Ennoblement and the Biofilm Analyses

The potential ennoblements observed at all stations and seasons were approximately equal, even though mass of biofilm formed on 654 SMO could differ by a factor of two decades between the different stations. We cannot explain the observed differences, but can speculate on possible causes such as:

(a) different hydrological conditions (see Tables 2 and 3).

(b) at Cherbourg and Kristineberg the sea water is circulating through large reservoirs offering sea water retention for an unknown period of time.

These may influence the rate of accumulation of biomass on the stainless steel surface.

The intracellular carbohydrate and protein content of the biofilm also showed a wide variation between both the different sites (Table 4) and different flow conditions (Table 5), while the extracellular polymer levels did not show significant differences between flow conditions (Table 5). These results are consistent with the variation in wet weight of the biofilm, since the intracellular carbohydrate and the protein content depend directly on the biomass. It was also possible to observe a broad relationship between the corrosion potential of the stainless steel and the quantity of EPS detected in the biofilm (see Figs 8 and 9). Again, this was not the case either for the intracellular carbohydrate content or for the protein content of the biofilm. These results indicate that the production of EPS on the metal surface during the development of the biofilm is a necessary condition for the potential ennoblement to occur. Despite the differences in the EPS content during the first days of exposure observed between the different marine stations (probably largely influenced by the conditioning of the metal surface before the biofilm growth) at electrode potentials below +150 mV SCE, a minimum quantity of EPS (approximatly 100 ng cm^{-2}) seems to be necessary in order to observe the ennoblement of the corrosion potential. These results are, to some extent, in agreement with that observed for biofilms formed on copper in potable water systems. In this case, it has been shown that EPS, through their cation selectivity, could be responsible for the microbially induced pitting corrosion of copper tubes [9]. However, even if the EPS content of the biofilm is an important factor, this may not be the only reason for the ennoblement of the corrosion potential. Thus, the amount of EPS may reflect the activity of the biofilm, through, for instance, the production of an oxidation agent by one or several enzymatic reactions, or to act as a trap for unknown biocatalysts [10]. More research is needed in order to define the mechanisms for the potential ennoblernent of stainless steel in natural sea water.

5. Conclusions

1. Ennoblement of the corrosion potential of 654 SMO was observed at all exposure sites and at all seasonal exposures.

2. No differences in the corrosion potential vs time curves were found for all investigated alloys at a given exposure site and season.

3. The ennoblement does not depend on the biomass of the biofilm.

4. The amount of EPS was more or less independent of the geographical location of the exposure site and the flow conditions.

5. A broad correlation has been established between the amount of EPS measured in the biofilm and the corrosion potential of stainless steel.

6. Acknowledgements

The authors would like to thank Avesta Sheffield (Sweden), Sandvik Steel (Sweden) and Creusot Loire Industrie (France) for their support and provision of the test materials. This study was supported in part by the Commission of the European Communities under Marine Science and Technology (MAST) contract MAS2-CT92-0011. The financial support from Nutek, Avesta Sheffield and Sandvik Steel for the Swedish Corrosion Institute is also gratefully acknowledged.

References

1. A. Mollica and A. Trevis, *Proc. 4th Int. Cong. on Marine Corrosion and Fouling*, Juan les Pins, France, 1976, p. 351.
2. R. Johnsen and E. Bardal, *Corrosion*, 1985, **41** (5), 296.
3. V. Scotto, R. D. Di Cientio and G. Marcenaro, *Corros. Sci.*, 1985, **25** (3), 185–194.
4. S. C. Dexter and G. Y. Gao, *Corrosion*, 1988, **44** (10), 717.
5. F. Mansfeld and B. Little, *Corros. Sci.*, 1991, **32** (3), 247–272.
6. F. Mansfeld, R. Tsai, H Shih, B. Little, R. Ray and P. Wagner, *Corros. Sci.*, 1992, **33** (3), 445–456.
7. P. Chandrasekaran and S. C. Dexter, *Corrosion '93*, Paper 493, NACE, Houston, Tx, 1993.
8. J. P. Audouard, C. Compère, N. J. E. Dowling, D. Féron, D. Festy, A. Mollica, T. Rogne, V. Scotto, U. Steinsmo, C. Taxen and D. Thierry, *Proc. 3rd Workshop on Microbial Corrosion*, Estoril, Portugal, 1994) European Federation of Corrosion publication No. 15, published by The Institute of Materials, London, 1995.
9. D. Wagner, W. R. Fisher and A. H. L. Chamberlain, *Proc. EUROCORR and UK Corrosion '94*, Bournemouth, UK. Published by The Institute of Materials, London, 1994.
10. V. Scotto, M. Beggiato, G. Marcenaro and R. Dellepiane, in European Federation of Corrosion Publication No. 10, pp. 21–33. Published by The Institute of Materials, London 1993.
11 . J. D. H. Strickland and P. R. Parsons, *A Practical Handbook of Sea water Analysis*. Fisheries Research Board of Canada, Bulletin 167, Ottawa. Canada, 1972.

3

Mechanism and Prevention of Biofilm Effects on Stainless Steel Corrosion

A. MOLLICA and V. SCOTTO

ICMM-CNR, via De Marini 6, 16149, Genova, Italy

ABSTRACT

The mechanism of oxygen reduction depolarisation induced by biofilm growth on stainless steels in sea water is discussed on the basis of the results of field tests performed within the MAST Contract No. MAS2 CT92 0011.

The data, both analytical and electrochemical, suggest that two concurrent oxygen reactions, with slow and fast kinetics respectively, take place on complementary areas of the metal surface. The extent of the θ fraction, where oxygen reduction is fast, depends on biofilm development as well as on the potential.

The application of intermittent or continuous chlorination as a prevention method against microbial corrosion induced by oxygen reduction depolarisation is also discussed.

The possible use of very simple electrochemical biofilm monitoring devices as a guide for intermittent chlorination in order to minimise biocide additions while, at the same time, maintaining their anticorrosive effect, is outlined.

It was finally shown that if chlorination is applied to reduce microbial induced corrosion, then heat exchange and wall smoothness will be retained, at the same time.

1. Introduction

It is now widely accepted that biofilm growth causes oxygen reduction depolarisation on active–passive alloys in sea water [1].

This phenomenon has attracted the attention of many people because of its likely severe consequences on the corrosion of many alloys used in industrial applications like stainless steels, Ti, Ni based alloys, and so on.

The diagram of Fig. 1 shows that oxygen reduction depolarisation induced by biofilm growth on active–passive alloys can:

- increase the likelihood of the onset of localised corrosion as a result of the ennoblement of the free corrosion potential of the alloy in the passive state,
- speed up the propagation rate of ongoing localised corrosion (active state),
- raise the galvanic current if these alloys are coupled with less noble alloys.

As a consequence, efforts have been made by several authors to quantify the corrosive effects of biofilm growth. An effort has also been made within the MAST Project, which, in particular, examined these effects on stainless steel (SS), considered as a typical example of an active–passive alloy.

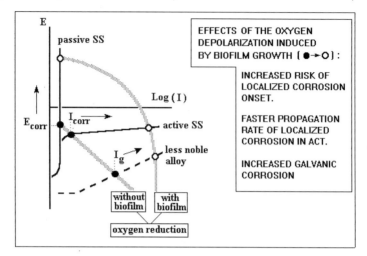

Fig. 1 *Effects of oxygen depolarisation induced by biofilm settlement on the corrosion of active–passive alloys and coupled less noble alloys.*

The common finding was that biofilm presence on exposed surfaces is a key factor in affecting the corrosion resistance of several alloys in sea water.

We can, therefore, proceed by examining two additional questions:

(a) Which is the mechanism by which biofilm growth causes oxygen depolarisation?
(b) How can the resulting corrosion be prevented?

2. The Mechanism of Oxygen Reduction Depolarisation Induced by Biofilm Growth on Stainless Steel Surfaces in Sea Water

The ennoblement of the free corrosion potential of stainless steel in the passive state as an effect of oxygen reduction depolarisation induced by bacterial settlement has attracted the attention of several authors. Many hypotheses have been proposed to explain this ennoblement, e.g. effects of metallorganic compounds, enzymes, peroxide, inhibitors, pH changes, etc. which are locally produced by bacteria [2].

Unfortunately, relatively few data have been provided to support the suggested theories.

Within the MAST project, an opposite approach has been adopted. Thus, rather than suggest a specific hypothesis, we have tried to provide field data so that future research can be conducted in promising directions.

With this objective in mind, we took two independent experimental approaches (described as Approach 1 and Approach 2), and then compared the end results.

2.1. Approach 1: Research of a Correlation between Change in Local Electrochemistry And Extracellular Products Buildup on Stainless Steel Surfaces

The research was based on the following considerations:

(i) The ennoblement of the free corrosion potential of stainless steel in the passive state occurs not only in sea water with very different salinity, temperature, pollution values [1] but also in estuarine [3] , fresh [4] and pond [5] water. It is unlikely that a single specific bacteria would be responsible for this phenomenon in all environments. For this reason, the study did not focus attention on the taxonomy of settled bacteria. Bacteria are more likely to act indirectly through the production of extracellular substances that can change the solution chemistry in close contact with the metallic surface and, hence, affect the local electrochemistry.

(ii) A preliminary attempt to search for a possible correlation between change in local electrochemistry and production of extracellular substances was made at ICMM. In this study, tests were conducted to identify any correlation between the evolution of the free corrosion potential of stainless steel in the passive state — which would indicate oxygen reduction depolarisation — and the build-up of extracellular carbohydrates (ESO) — which would indicate extracellular production by settled bacteria. The study has specifically examined only the range of free corrosion potentials over +100 mV (SCE).

The result of a previous test [6] was encouraging, showing a linear correlation between the two parameters. In agreement with our MAST project partners, it was decided to check the validity of this approach on a European level.

All partners exposed SS samples (in the form of plates and tubes) at their respective sea stations, with tests repeated in spring, summer, autumn and winter. Biofilm samples were collected during the period of development of the free corrosion potential in each exposure and the biofilm components extracted and finally sent to ICMM for analysis.

Some data scatter, due to different handling procedures by each partner during biofilm collection and extraction, was to be expected. The graph in Fig. 2 shows the results of this joint study [7], and a few comments can be made about the results obtained:

 • The hatched area contains *ca.* 80% of all the data, although obtained under very different SS exposure conditions in sea water. In the same graph all the data obtained during the exposure of different quality SS, in the temperature range of 6–28 °C, with a sea water velocity range between almost 0 and about 1 ms^{-1}, are plotted together.

  Under the conditions of the tests, the time required for biofilm development and also the overall final biomasses were very different. However, when the biofilm was analysed in term of ESO, the same relationship linking the corrosion potential E_{corr} and ESO evolution was obtained.This suggests that

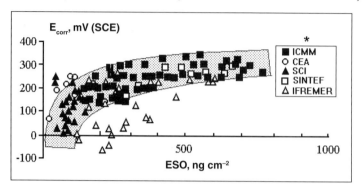

Fig. 2 *Experimental correlation between free corrosion potential of SS in passive state and extracellular carbohydrate (ESO) buildup on samples exposed to natural sea water [7].*

there is a close link between oxygen reduction depolarisation and extracellular product buildup on SS surfaces, irrespective of the exposure conditions in which this buildup takes place.

- On the other hand, the graph shows that the relationship between the two tested parameters is not really simple. A sharp increase in free corrosion potential up to +100 mV(SCE) is observed corresponding to the first 150 ng cm^{-2} of ESO buildup on the SS surfaces. A further ESO buildup up to *ca.* 1000 ng cm^{-2} slows down the potential increase. In this latter phase, an almost linear correlation between the two parameters is observed, which is very similar to that previously obtained in Genoa [6] in the same potential range.

   A mechanistic interpretation to justify this experimental trend is required.

2.2. Approach 2: Research of an Analytical Description of Oxygen Reduction Evolution During Biofilm Growth

Figure 3 shows the evolution of the oxygen reduction current on five SS samples concurrently exposed to natural sea water and polarised at 150, 0, –150, –300 and –450 mv (SCE) respectively.

   The cathodic current at each imposed potential was automatically measured and recorded every 6 h by a data logger. Initially clean SS surfaces were colonised by bacteria during their exposure in sea water.

   It can be seen that, at the beginning of the exposure, the measured oxygen reduction currents can be covered by the line 1, whose equation is

$$i_1(E) = 10^{-8.3} \; 10^{-E / \beta_1}$$

where $\beta_1 = 0.23$ V/decade. Several experimental points fall near this line suggesting that, for a few days, the oxygen reduction kinetics did not change. Later, the sequence of measured points indicates that, at each imposed potential, the cathodic current increased rapidly for 2–3 d and, finally, remained stable.

Curve 2, which covers the final points, shows that oxygen reduction was strongly depolarised: a limiting reduction current can even be observed at a potential close to −100 mV(SCE).

An attempt was made to find an analytical expression that would describe this evolution; for example an expression like $i = f$ (imposed potential, biofilm evolution).

The first attempt at producing this expression was based on the following considerations:

– only two stable oxygen reduction kinetics can be observed: the first $[i_1(E)]$, slow, in absence of biofilm, the second $[i_2(E)]$, fast, when the biofilm is completely developed on s.s.surfaces;

– we can, therefore, surmise that the evolution of cathodic current with time results from the concurrent evolution of two fixed oxygen reduction kinetics, $i_1(E)$ and $i_2(E)$, respectively taking place on complementary surface area ratios $[(1−\theta(t))$ and $\theta(t)]$; $\theta(t)$, which increases with time, is the overall extent of "active sites", generated by biofilm growth, where oxygen reduction is fast.

It follows that:

$$i(E,t) = i_1(E)\,(1−\theta(t)) + i_2(E)\,\theta(t) = i_1(E) + [i_2(E) − i_1(E)]\,\theta(t) \qquad 0 \le \theta(t) \le 1. \qquad (1)$$

– Finally we can assume that the extent of "$\theta(t)$ sites" is proportional to the bacterial population settled on SS surfaces and, as a consequence, its evolution with time

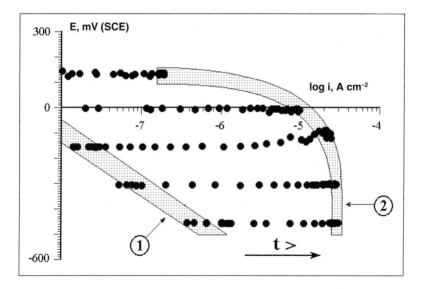

Fig. 3 *Oxygen reduction current evolution on SS surfaces which are continuously cathodically polarised at fixed potentials during the exposure to natural sea water [8].*

follows the same logistic expression that is often used to describe biofilm growth [9]. The following expression was assumed for $\theta(t)$:

$$\theta(t) = [0.1 \bullet 10^{(t-t_o)/\tau} / (0.9 + -0.1 \bullet 10^{(t-t_o)/\tau})] \tag{2}$$

where t_o and τ represent, respectively, bacterial population incubation and duplication time.

Equations (1) and (2) were used to fit the experimental evolution of the cathodic current in order to check the validity of the proposed calculation procedure.

In Fig. 4 the same data already plotted in Fig. 3 are reported in a different form to show more clearly the cathodic current time dependence at each imposed potential. The continuous lines represent the trends calculated through eqns (1) and (2), for the set of values reported in the Figure.

The comparison between the experimental and calculated trends show that:

– at each potential, a set of values for i_1, i_2, t_o and τ can be found in order to properly fit the real current evolution using a logistic expression for $\theta(t)$;
– furthermore, in the potential region below –150 mV(SCE), the value of the two t_o and τ time constants are very close together, and, in addition, close to the values obtained in previous tests, when a logistic expression was applied to fit the results of settled bacteria counts. A close correlation between $\theta(t)$ evolution and biofilm growth is then confirmed in the potential region below –150 mV(SCE).
– on the other hand, since the specimens were put in the same water simultaneously, a single biofilm evolution law, and therefore a single couple of t_o and τ values, should be expected on all tested samples. On the contrary, both t_o and τ, obtained by fitting the i–t curves, are clearly increasing regularly when the imposed

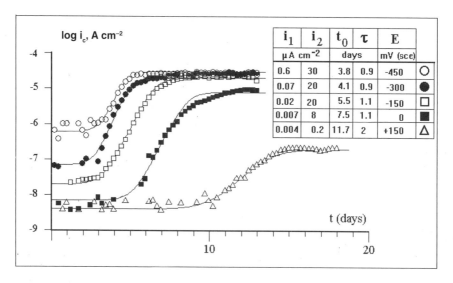

i_1	i_2	t_0	τ	E	
μA cm^{-2}		days		mV (sce)	
0.6	30	3.8	0.9	-450	○
0.07	20	4.1	0.9	-300	●
0.02	20	5.5	1.1	-150	□
0.007	8	7.5	1.1	0	■
0.004	0.2	11.7	2	+150	△

Fig. 4 Fitting, through eqns (1) and (2), of the experimental data already plotted in Fig. 3 [8] .

potential is raised above −150 mV(SCE).

We concluded that a correction must be applied to eqn (2) to extend its validity in the region of relatively noble potentials.

Equation (2) was, then, modified as follows:

$$\theta(t,E) = [0.1 \bullet 10^{(t-t_o)/\tau} / (0.9 + \; -0.1 \bullet 10^{(t-t_o)/\tau})] \bullet [k\exp(-E/\beta)/(1+k\exp(-E/\beta))]$$

$$(3)$$

by adding a correction which is relevant only at relatively high potentials, since it tends towards 1 for low potentials.

Figure 5(a) shows the expected current evolution calculated every 6 h by eqn (1) using the expression (3) to calculate θ and assuming β = 0.15 V/decade, k = 0.4, t_o= 5 d and τ =1 d.

By comparing the data of Fig. 5(a) with the data plotted in Fig. 3 a sufficient agreement can be observed between the calculated and experimental data at all imposed potentials.

Finally, Fig. 5(b) shows the calculated trend of $\theta(t,E)$: a sharp decrease in the extent of active sites is expected to occur when the potential is increased above −150 mV (SCE).

In summary, the analysis of the experimental current evolution on SS exposed to natural sea water suggests that:

– two concurrent oxygen reduction kinetics, i.e. slow and fast, can occur at the same time on complementary surface areas;
– the θ extent of the surface ratio, where oxygen reduction is fast, depends both on biofilm growth and on imposed potential.

Preliminary tests conducted at the Swedish Corrosion Institute (SCI) [10], where cathodic current distribution on SS samples exposed to natural sea water was examined through surface inspection by Scanning Vibrating Electrode Techniques (SVET), substantiated both the above conclusions.

2.3. Consistency of Results of the Two Approaches

Equation (1), in which θ was calculated from eqn (3), was sufficiently accurate to describe oxygen reduction current evolution at fixed potentials (potentiostatic mode).

This equation can be rearranged to anticipate the potential evolution at a fixed cathodic current (intensiostatic mode) during biofilm development on an SS surface.

If, in an intensiostatic mode, the cathodic current is set equal to the s.s.passivity current in the passive state, then the rearranged equation can be used to calculate the free corrosion potential evolution vs biofilm growth.

To conduct this calculation, a cathodic current value close to 10^{-8} A cm^{-2} was selected and, using the data plotted in Fig. 2, the complete biofilm development was normalised to an ESO amount equal to 1000 ng cm^{-2} (Fig. 6).

In agreement with the experimental pattern (Fig. 2), the calculated E_{corr} vs ESO

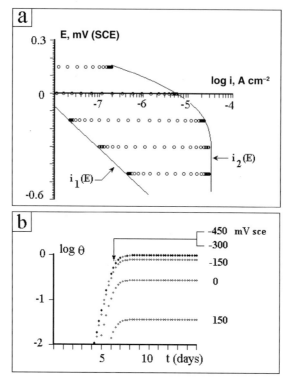

Fig. 5 *Evolution of (a) oxygen reduction current, and (b) of θ during SS exposure to natural sea water calculated through eqns (1) and (3), every 6 h, at different potentials [8].*

curve shows that a fast potential increase must be expected during the initial ESO buildup to an amount close to 150 ng cm^{-2}, followed by a slower potential increase which will be linearly dependent on additional ESO buildup.

Therefore, the two independent experimental approaches, the analytical and the electrochemical, undertaken to consider promising directions for future research work on the mechanism of biofilm effect on the corrosion of active–passive alloys, are consistent.

Everything suggests that settled bacteria produce some extracellular subtances that, if adsorbed on the metallic surface, act as a catalyst for oxygen reduction. The amount of adsorbed catalyst depends both on biofilm growth and on the potential.

2.4. Which Catalyst?

The experimental data in Fig. 2, and those calculated in Fig. 6, clearly suggest that future research must focus on extracellular production but cannot help in the identification of the oxygen reduction catalyst for which we are looking.

Obviously, a curve equal to that of Fig. 2, (except for the *x*–axis units), will be obtained if, instead of ESO, the biofilm is expressed in terms of whatever other substance, proportional to ESO, is produced by the bacteria.

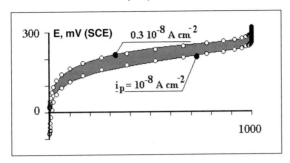

Fig. 6 *Expected evolution of free corrosion potential of SS in passive state vs ESO, calculated by rearrangement of eqn (3).*

We have, nevertheless, obtained some additional information which can help future research.

As already proved [11–13] and confirmed in the MAST project by SCI and the Commissariat à l'Energie Atomique (France), CEA [10], we are looking for a substance which can be inactivated either by adding an enzyme inhibitor (such as sodium azide) to the sea water or by increasing the sea water temperature to 40°C.

3. Protection of Stainless Steel against Microbially Induced Corrosion

To the question "How can SS be protected against the corrosive effect of biofilm growth?", there is a straightforward answer: "Biofilm formation must be prevented".

In industrial plants the fight against biofilm, which is normally conducted to improve heat exchange and reduce the friction factor, is usually performed by chlorination.

On the one hand, biocide addition causes a reduction in MIC, but on the other hand the biocide normally applied is an oxidising agent which is itself corrosive.

The question then arises as to when the anticorrosive effect (MIC reduction) of the antifouling treatment prevails over the corrosive action of the oxidant.

Information is required on the correct treatment in terms of application frequency, biocide concentration and treatment duration in order to produce a prevailing anticorrosive effect.

Similarly, it is important to know whether chlorination can be optimised through the use of alternative biofilm monitoring devices.

Finally, if corrosion is prevented through a suitable antifouling treatment, is the heat exchange improved and the friction factor reduced at the same time?

Studies on these, and similar problems, were planned as part of the MAST project.

To deal with the relations between biofilm, chlorination and corrosion, we shall first examine continuous chlorination and, then, intermittent chlorination.

3.1. Continuous Chlorination

The graph in Fig. 7, which summarises the results of several field tests, provides an overview of chlorination effects on SS corrosion [14].

3.1.1. Effect of continuous chlorination on the likelihood of the onset of localised corrosion
Taking the level of the maximum free corrosion potential of SS in the passive state as an index of the risk of the onset of localised corrosion, the graph suggests that with an increase in residual chlorine concentration (rcc) above approx. 0.2 ppm, this risk is higher than that caused by biofilm growth in untreated sea water.
   In more detail, the graph in Fig. 8 shows that the minimum free corrosion potential is obtained in case of continuous chlorination corresponding to 0.1 rcc. Biofilm analyses showed that this rcc is just enough to avoid biofilm growth.
   If rcc is increased over this optimum value, the free corrosion potential rapidly rises over the value induced by biofilm growth in untreated sea water.
   It follows that the risk of the onset of localised corrosion can be minimised by continuous chlorination, provided rcc is properly controlled within a very narrow range close to 0.1 ppm. Otherwise, the risk is increased.

3.1.2. Effect of continuous chlorination on ongoing localised corrosion propagation
From Fig. 7, it is seen that, whatever the chlorination level, the propagation rate of ongoing localised corrosion (SS in active state) is always reduced. The minimum corrosion rate corresponds to the chlorination level which is just enough to avoid biofilm growth on the surrounding SS cathodic surfaces (0.1–0.2 ppm rcc).

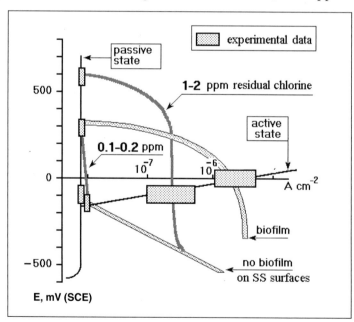

Fig. 7 *SS behaviour in untreated and chlorinated sea water [14].*

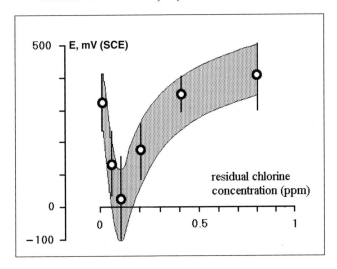

Fig. 8 Free corrosion potential of SS in passive state after 30 d of exposure to continuously chlorinated natural sea water [15].

Figure 9 provides direct evidence of the effect of chlorination on crevice corrosion propagation with AISI 316 type stainless steel.

3.1.3. Effect of continuous chlorination on galvanic currents
Obviously, the effect of chlorination on galvanic currents will be the same as that observed for localised corrosion propagation: since the pattern is essentially the same.

The graph in Fig. 10 shows, in particular, the effect of chlorination on the galvanic couple between a SS pipe and an iron sacrificial anode. It can be seen that, when biofilm growth is prevented, the galvanic current decreases by about one order of magnitude and is almost insensitive to rcc fluctuations between 0.1 and 0.8 ppm.

Therefore, if compared with untreated sea water, continuous chlorination acts as (i) a promoter of the initiation of localised corrosion if the rcc exceeds a narrow range close to 0.1 ppm, and (ii) an inhibitor both of the propagation of localised corrosion and of the galvanic corrosion irrespective of the chlorination level (at least up to 10 ppm rcc).

3.1.4. Protection against the onset of localised corrosion
The above statements can be viewed in a different way; one which is of greater use for practical applications:

• If the galvanic current is strongly reduced, whatever the rcc, then a weak cathodic protection can be applied to protect the SS against the risk of the onset of localised corrosion, without particular attention to chlorination dose, thus avoiding the only risk induced by chlorination.

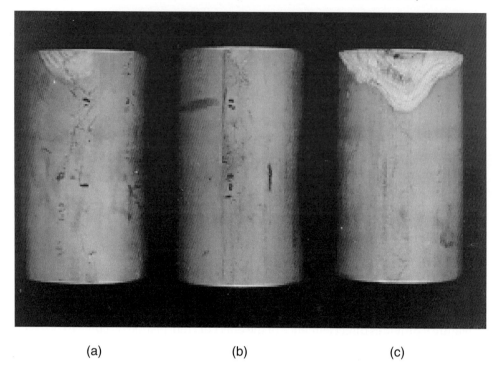

(a)   (b)   (c)

Fig. 9 Crevice corrosion attack on AISI 316 after 75 d of exposure [14] to (a) continuously chlorinated sea water at 1–2 ppm residual chlorine concentration (rcc); (b) continuously chlorinated sea water at 0.1–0.2 ppm rcc, and (c) untreated sea water.

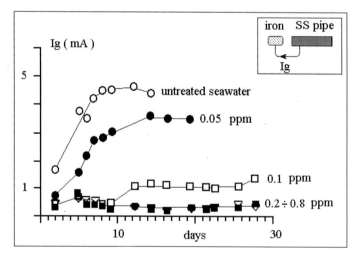

Fig. 10 *Galvanic current trend between SS tubes and Fe anodes in continuously chlorinated sea water at different residual chlorine concentration [15].*

• In addition, if the protection current is low and, as a consequence, the ohmic drop in solution is negligible, protection against the risk of the onset of localised corrosion in chlorinated sea water can be applied even in difficult geometric configurations. Figure 11 shows an example of this protection system applied to a SS pipe. It can be seen that in untreated sea water, the potential inside a SS pipe joined to an iron sacrificial anode rapidly rises owing to the high ohmic drop caused by the fast oxygen reduction current on fouled SS surfaces. When biofilm development is prevented by chlorination above 0.1 rcc, the protection current is reduced by one order of magnitude and the potential profile remains flat for several metres, at a value which is highly protective against localised corrosion onset on SS, whatever the rcc.

An interesting application of inexpensive cathodic protection on SS pipes exposed to chlorinated sea water was studied at SINTEF [16].

In conclusion, a combined chlorination–cathodic protection system seems to be the best solution against all the corrosive effects of biofilm growth on SS surfaces.

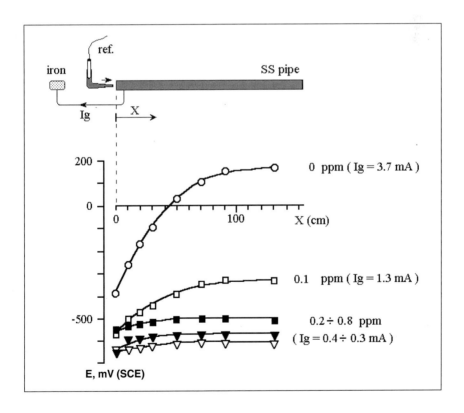

Fig. 11 *Potential profile inside SS pipe cathodically protected by an iron anode after 30 d exposure to continuously chlorinated natural sea water depending on residual chlorine concentration; the corresponding galvanic current values are reported in brackets [15].*

3.2. Intermittent Chlorination

3.2.1. Parameters for intermittent chlorination
The graph in Fig. 12 shows an example of the effect of intermittent chlorination on galvanic currents. During the treatment, residual chlorine concentrations ranging from 0.2 to 1.5 ppm were applied [17].

The Figure shows that the galvanic current increased as a result of biofilm growth on SS surfaces, but could be brought back to the initial value when proper chlorination was applied. This procedure could be repeated after biofilm regrowth.

A careful examination of the data indicates that:

–   a fast current decrease was obtained if the rcc (C) was at least greater than 0.4 ppm;
–   for rcc (C) exceeding 0.4 ppm, current decay during chlorination followed the equation

$$i = i_{min} + (i_{max} - i_{min}) \bullet 10^{(-\alpha C t)} \tag{4}$$

where the product $C.t = Q$ represents the total amount of biocide added to the sea water.

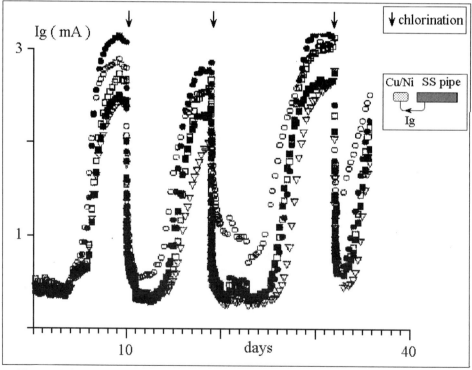

Fig. 12 *Galvanic current trend between SS tubes and iron anodes exposed to intermittently chlorinated sea water. Tested residual chlorine concentrations ranged between 0.2–1.5 ppm [17].*

We conclude that suitable intermittent chlorination can effectively act as a temporary corrosion inhibitor of galvanic corrosion (and also of localised corrosion propagation) if a suitable quantity of biocide is added to the solution and the concentration maintained above a minimum level.

Obviously, the time averaged corrosion inhibition efficacy of intermittent chlorination will depend on the frequency of biocide application.

It is also clear from Fig. 12 that if chlorination is significantly delayed after biofilm growth, the average galvanic current will be high. On the other hand, treatment during the incubation phase or before biofilm regrowth is useless.

3.2.2. Optimisation of chlorination against MIC, in terms of application frequency and amount of biocide addition in sea water, through the use of an electrochemical biofilm monitoring device

To optimise biocide application, the following information is required:

- in order to minimise biocide addition frequency we need to know when the settled biofilm starts to be corrosive;
- in order to minimise the biocide amount added to sea water during chlorination, we need to know if the biocide concentration multiplied by the treatment duration is sufficient to remove the corrosive effects of a biofilm that has already settled.

The above information can hardly be obtained *a priori* through models, because biofilm growth and its destruction depend on many factors, such as sea water temperature, composition, flow rate, previous treatments and so on.

Monitoring devices capable of real time and *in situ* evaluation of biofilm corrosivity could provide the required information.

The data plotted in Fig. 12 suggest a very simple way to make an electrochemical biofilm monitoring device.

In fact, the time dependence of the galvanic current in a simple galvanic couple, such as that shown in the Figure, can provide the requested information.

We carried out several field tests using such a device as a guide for automatic chlorination.

Chlorination was automatically applied when the signal sent by the device inserted in an experimental loop (see Fig. 13) exceeded a preset "risk threshold" level, as a consequence of biofilm growth on the SS surfaces. It was automatically stopped when the signal fell below a second prefixed level, following biofilm destruction. Biofilm regrowth would prompt a new chlorination, applied according to the same rule.

Field test results [10] are summarised in the following section.

3.2.3. Minimised MIC

Figure14 shows, as an example, the output of two electrochemical biofilm monitoring devices, exposed to automatically chlorinated and untreated sea water respectively, during a test performed in summer.

It is clear that automatic chlorination led to a strong reduction in average galvanic

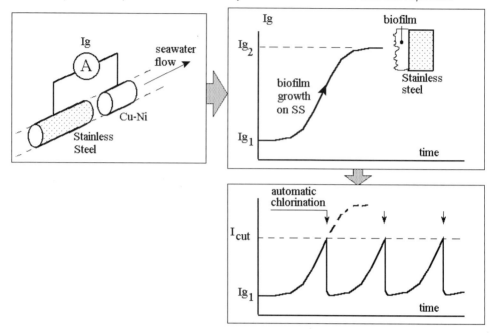

Fig. 13 *Possible application of an electrochemical biofilm monitoring device as guide for automatic chlorination.*

corrosion value and the same strong inhibitive effect was also observed on crevice corrosion propagation.

As shown in Fig. 15, after 3 months of exposure in untreated sea water, the AISI316 pipes were perforated by crevice corrosion, whereas similar SS pipes exposed to automatically chlorinated sea water suffered only a superficial attack. As described above, this residual corrosion could be easily avoided through weak cathodic protection.

3.2.4. Minimised chlorination frequency

Studying the behaviour of an electrochemical biofilm monitoring device as a guide for automatic chlorination, we conducted several field tests covering one year of exposure to natural sea water (Fig. 16). During this period, the sea water temperature ranged from 28–29°C to 14°C.

The automatic chlorination frequency that is required strongly depends on sea water temperature; according to the sensor, no more than one chlorination every 6 d is required in winter to protect it against microbiological induced corrosion.

In other words, the sensor is able to follow the change in biological activity induced by varying environmental parameters and automatically adjusts the chlorination frequency.

3.2.5. Concurrently minimised friction factor and thermal exchange resistance

The photograph in Fig. 17 shows the appearance of two SS tubes exposed for three

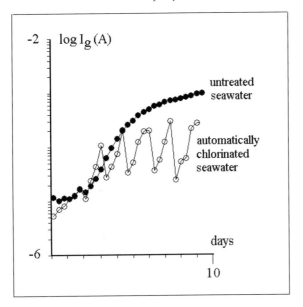

Fig. 14 *Output of two sensors exposed to untreated and automatically chlorinated sea water, respectively [10].*

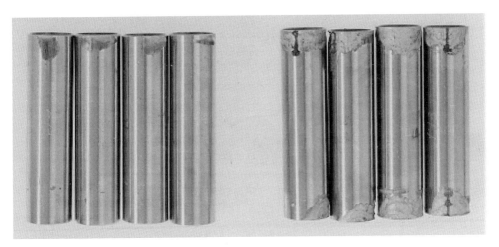

Fig. 15 *Crevice corrosion propagation on AISI 316 pipes after three months of exposure to untreated (right side) or to automatically chlorinated (left side) sea water.*

months to untreated and automatically treated sea water, respectively: the first is heavily fouled, whereas the second appears completely clean.

EDAX analyses show that barely the peaks of the base alloy elements (Fe, Cr, Ni, Mo) can be observed on the surfaces exposed to automatically treated sea water. In

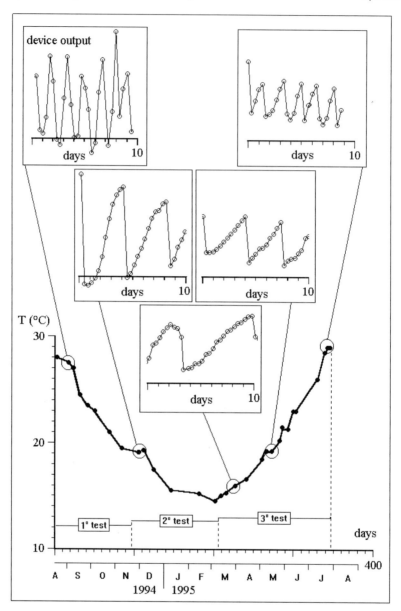

Fig. 16 *Relation between automatic chlorination frequency and sea water temperature [10].*

untreated sea water, Si, and Al peaks almost cover base alloy peaks showing that exposed SS surfaces are covered by silt and clay trapped in biofilm mucilage.

It follows that, if MIC is controlled through intermittent chlorination automatically guided by an electrochemical biofilm monitoring device, both wall smoothness and thermal exchange will be preserved.

This result is due to the fact that an electrochemical biofilm monitoring device

Fig. 17 *Appearance of internal pipe walls after 3 months of exposure to untreated (left) and automatically chlorinated sea water (right) [18].*

acts as a very early warning system of biofilm growth, since it is able to signal biofilm presence long before plant efficiency is affected.

In practice, only few bacteria can be detected by SEM observation on SS surfaces exposed to automatically chlorinated sea water, whereas a mature biofilm appears on the surfaces exposed to untreated sea water (Fig. 18).

In agreement with these conclusions, field tests performed at IFREMER [10] showed that an electrochemical biofilm monitoring device is able to signal biofilm presence long before any conventional biofilm monitoring device based on heat transfer measurements.

4. Conclusions

With regard to the mechanism of oxygen reduction depolarisation induced by biofilm growth on stainless steels exposed to natural sea water, field data, both analytical and electrochemical, obtained by the MAST Project suggest the following:

1. Two oxygen reduction kinetics, i.e. slow and fast respectively, can concurrently take place on complementary areas of a stainless steel surface immersed in sea water;
2. Fast kinetics are due to the adsorption of some extracellular bacterial product, which acts as an oxygen reduction catalyst and can be inactivated by an enzyme inhibitor such as sodium azide or by a temperature increase to 40°C;

3. The extent of the fraction of the surface area, on which the fast oxygen reduction occurs, depends both on biofilm development and on the potential.

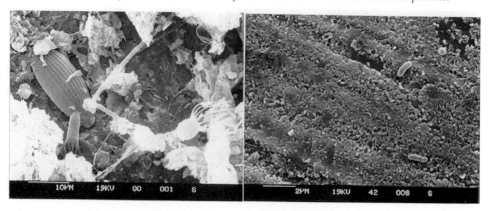

Fig. 18 *Scanning electron micrographs of SS pipe walls after three months of exposure to untreated (left) and automatically chlorinated sea water (right) [18].*

With regard to the prevention of microbial corrosion induced by oxygen reduction depolarisation, the use of chlorination, both continuous and intermittent, as a preventive method against MIC, data suggest the following:

4. Continuous chlorination at residual chlorine concentration (rcc) close to 0.1 ppm prevents biofilm settlement and reduces the risk of the onset of localised corrosion compared with untreated sea water;

5. Continuous chlorination at rcc above 0.2–0.3 ppm greatly increases the risk of the onset of localised corrosion. This adverse effect can be offset with the application of a weak cathodic protection current of 10^{-7} A cm^{-2}, as a maximum order of magnitude;

6. Continuous chlorination at more than 0.1 ppm residual chlorine concentration, or intermittent chlorination at more than 0.4 ppm act, in any case, as inhibitors of ongoing localised corrosion of stainless steels as well as of galvanic corrosion induced on less noble materials coupled to SS;

7. Very simple electrochemical biofilm monitoring devices can provide useful real time information to guide intermittent chlorination in order to minimise biocide addition while, at the same time, maintaining an anticorrosive effect;

8. When chlorination reduces microbially induced corrosion, the biofilm thickness will be controlled at a microscopic level. Control of corrosion by this method will also maintain heat exchange and wall smoothness.

References

1. A. Mollica, *Int. Biodeterior. & Biodegrad.*, 1992, **29**, 213.
2. P. Chandrasekaran and S. C. Dexter, *Corrosion '93*, Paper 493.
3. S. C. Dexter and H. J. Zhang, *Proc. 11th Int. Corrosion Congr.*, Florence, 1990, Associazione Italiana Di Metallurgia, Italy, 4, p.333.
4. S. C. Dexter and H. J. Zhang, Project 2939–4. Final Report, University of Delaware.
5. S. Maruthamuthu, G. Rajagopal, S. Sathianarayannan, M. Eashwar and K.Balakrishnan, *Biofouling*, 1995, **8**, 223.
6. V. Scotto, M. Beggiato, G. Marcenaro and R. Dellepiane, Working Party Report on "Marine Corrosion of stainless steels: Chlorination and Microbial Effects" Publication No.10 in European Federation of Corrosion series. Published by The Institute of Materials, London, 1993, p.149.
7. J. P. Audouard, C. Compère, N. J. E. Dowling, D. Féron, D. Festy, A. Mollica, T. Rogne, V. Scotto, U. Steinsmno, C. Taxen and D. Thierry, *Proc Int. Congr. on Microbial Induced Corrosion*, New Orleans, USA, May 1995.
8. A. Mollica and E. Traverso, *Proc. of Int. Conf. on Corrosion in Natural and Industrial Environments: Problems and Solutions*, Grado 1995, NACE Inter. Italia Section, p.249.
9. W. G. Characklis, Kinetics of microbial transformation, in *Biofilm*, (W. G. Characklis and K. C. Marshall, eds), 1990, p.233.
10. MAS2–CTT92–0011 Report on *Proc. of 2nd MAST Days and EUROMAR Market*, Sorrento, Italy, 7–10 November 1995, p.1059.
11. V. Scotto, R. DiCintio and G. Marcenaro, *Corros. Sci.*, 1985, **25**, 185.
12. R. Johnsen and E. Bårdal, *Corrosion*, 1985, **41**, 296.
13. A. Mollica, A. Trevis, E. Traverso, G. Ventura, G. DeCarolis and R. Dellepiane, *Corrosion*, 1989, **45**, 48.
14. G. Ventura, E. Traverso and A. Mollica, *Corrosion*, 1989, **45**, 319.
15. A. Mollica, E. Traverso and G. Ventura, *Proc. 11th Int. Corrosion Congr.*, Florence, 1990, Associazione Italiana Di Metallurgia, Italy, 4, p.341.
16. J. M. Drugli *et al.*, This publication p.165.
17. A. Mollica and G. Ventura, *Proc. 8th Int Congress on Marine Corrosion and Fouling,* Taranto, (1992), Italy, in *Oebalia*, 29, (suppl. 1993), p.313.
18. A. Mollica and G. Ventura, *Proc. 12th Int. Corrosion Congress*, 1993, NACE, Houston, Tx, USA, 1993, Vol 5B, p.3807.

4

Biocide Options in the Offshore Industry

P. CARRERA and G. GABETTA

Eniricerche S.p.A. S. Donato Milanese, Italy

ABSTRACT

The main factors which make sea water a very corrosive fluid can be divided into two groups: (bio)chemical, (such as oxygen, carbonate, salts, organic compounds, biochemical activity and pollutants), and physical (temperature, flow velocity, potential, pressure and light).

Microbiologically Influenced Corrosion (MIC) in stainless steels acts usually via the depolarisation of the oxygen reduction process. The presence of a bacterial mass implies the production of extracellular polymeric substances (EPS) and bioproducts, which act as catalyst of the oxygen reduction.

In offshore systems, stainless steels are mainly used for hydraulic sea water handling systems. To avoid fouling, the water is usually treated with biocides (mostly chlorine), using intermittent or low-level continous regimes. The effect of chlorine concentration is of great importance for both chemical and biological corrosion: if chlorine is used for bacterial control and not added within tight limits it can promote localised corrosion. The antifouling procedures based on biocides often cause additional problems related to environmental impacts, law restrictions and safety.

In this paper, a short review is presented of the different biocide options more commonly used in the offshore industry for the antifouling of stainless steel piping and components.

1. Introduction

Corrosion in sea water is caused by electrochemical factors as well as the formation of biofilm. Biofilm forms on the surface of components in contact with water and accumulates in a non-uniform way, growing in time and in space [1]. The biofilm formation inside pipelines is a function of water flow rate; depending on the bacterial species involved, aerobic biofilm can be observed (usually in the external part of the deposit), while anaerobic bacteria are generally present in the internal part, close to the substrate.

Biofilms on the metal substrate can influence corrosion mechanisms, as a result of the formation of occluded areas, as well as biological processes. In the case of stainless steels in sea water, experimental evidence [2] suggests that a connection exists between biofilm growth and the observed oxygen reduction depolarisation, as schematically indicated in Fig. 1.

The control of biofilm is crucial when attempting to limit corrosion problems in those components. For an effective control, physical and mechanical changes should be coupled with an appropriate biocide treatment [3]. The selection of the chemical

An hypothesis on marine biocorrosion mechanism

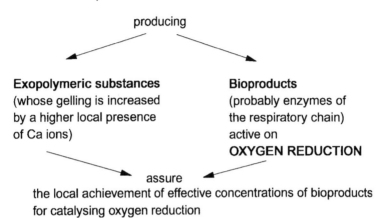

A bacterial presence of 10^8 cells cm^{-2} at least

producing

Exopolymeric substances
(whose gelling is increased
by a higher local presence
of Ca ions)

Bioproducts
(probably enzymes of
the respiratory chain)
active on
OXYGEN REDUCTION

assure
the local achievement of effective concentrations of bioproducts
for catalysing oxygen reduction

Fig. 1 *Scheme of the MIC mechanism on stainless steel in sea water [2].*

product depends on the characteristics of each individual site. The type of biocide to be used and the treatment regime are interdependent. The treatment can be batch (or "slug"), continuous, or a combination of the two. Biocides can also be released slowly by a painting or coating, as in the case of ships or offshore structural components [4, 5].

Some factors to be considered when selecting the most appropriate treatment methods are:

1. Economic consideration;
2. Distribution of bacterial contamination;
3. Treatment facilities and location of the component to be treated;
4. Climate or weather;
5. Environmental problems and water quality;
6. Mutual influence with other existing chemical treatment

Antifouling procedures, mainly based on the use of biocides, often give rise to problems related to environmental impact, law restriction and safety. To minimise pollution, the quantity of added chemicals must be kept to a minimum. It is important, therefore, to develop means of detecting negative biological effects (on both micro- and macro-organisms) as early as possible, and of monitoring bacterial activity (before, during and after) the biocide treatment in order to control and implement its efficacy. A better understanding of active mechanisms would be helpful.

When biocides are components of paints and coatings, the active chemicals are slowly released by the paints into the sea water. A more conscious aptitude toward environmental issues is also necessary when using naval paints. An innovative approach to the problem is the study of antifouling substances produced by marine organisms, such as algae, fungi, corals and vertebrates. These natural compounds would be able to control the biofilm without damage to the environment [6].

2. Oil Production Systems

In oil production systems, stainless steels are used for many components, such as those listed in Table 1 [7]. The sea water handling systems on offshore platforms supply water to crude oil and gas coolers, drilling operations, downhole injection and storage ballast, fire main systems and living quarters [8].

Possible problems are related to the presence of chlorine and chloride ions in the water, aerobic and anaerobic MIC, bacteria and algae. Local temperature increases can cause slime formation, localised attack and so on. Countermeasures are physical cleaning (use of an appropriate pig or water flush system, wash with fresh water during stops, backflow of the injection well, acidising) and mechanical changes (to mantain local water flow rate, eliminate dead spots, reduce water handling time, remove oxygen from the system, etc.). Water treatment and /or biocide injection are widely used.

A list of components susceptible to MIC problems in oil plants and offshore structures is summarised in Table 2 [3].

3. Types of Biocides

Biocides can be broadly classified into two groups: oxidising and non-oxidising biocides [9].

Table 1. Examples of stainless steel in oil production systems [7]

Field	Operated by	System	Pump Material
Gullfaks	Statoil	Cladding in pressure wells	AISI 316L
		Produced water	6Mo-steel
Tommeliten	Statoil	9" (23cm) flowline	Duplex stainless steel (UNS S31803)
Snorre	Saga Petroleum	Process piping	6Mo-steel
Oseberg	Norsk Hydro	Process piping	Duplex stainless steel (UNS S31803)
Veslefrikk	Statoil	Process piping	Duplex stainless steel (UNS S32760)

Table 2. MIC susceptible components in oil production plants

Component	Location	Type of bacteria	Countermeasures
Flow lines	Stagnant points – low flow rate	Sulphate Reducing Bacteria (SRB), Slime Forming Bacteria (SFB)	Clean by pigging; Increase flow rate, avoid dead spots; Biocides
Washtanks	Underscale deposits	SRB	Clean, acidise; Biocides
Storage tanks	Under scale deposits	SRB, SFB	Clean; Stir; Biocides
Bottom of pits	Under sludge or in the mud	SRB	Biocides
Filters		SRB, SFB, Iron Oxidising Bacteria (IOB)	Back-wash; Biocides
Rat holes	In the fluid	SRB	Biocides
Annular space	In the packer fluid	SRB	Biocides
Casing	Insoil and drilling mud	SRB	Adjust Cathodic Protection
Heaters-treaters	Oil–water interface	SRB, SFB	Biocides
Buried pipelines	Outside, in the soil	SRB	Adjust Cathodic Protection
Storage legs	Dead spots	SRB	Eliminate dead spots; Biocides
Water injection system	Pumps, reservoir, deaeration system, heat exchangers	SRB	Segregate fresh and sea water when possible; Biocides
Open ponds	Edge of the ponds	SFB, IOB	
Supply wells	Oxygenated water	SFB, IOB	Deoxygenate; Avoid flow changes
Injection wells		SFB, IOB	Back flow, acidise, avoid flow changes; Biocides

Listed below are the most commonly used oxidising biocides:

1 Chlorine (Cl_2)
2 Chlorinating compounds [$NaOCl$, $Ca(OCl)_2$]
3 Chlorine dioxide (ClO_2)
4 Chloramines (Ammonium Chloride, [NH_2Cl])
5 Bromine (Br_2).

The non-oxidising biocides include:

- Aldehydes
  →Formaldehyde (HCHO)
  →Glutaraldehyde (pentanedial, $OCH[CH_2]_3 CHO$)
  →Acrolein ($H_2C=CHCHO$; 2 propenal)

- Amine-type compounds
  →Quaternary amine compounds
  →Amine and diamine

- Halogenated compounds
  →Bronopol ($CH_3H_6BrNO_4$)
  →DBNPA (2,2-dibromo-3-nitrilo-propionamide; $CHBr_2 CNCONH_2$)

- Sulfur compounds
  →Isothiazolone (C_4H_4NOClS)
  →Carbamates
  →Metronidazole (2-methyl-5-nitroimidazole-1-ethanol)

- Quaternary phosphonium salts.

Biocides are difficult to select because they are used to combat a problem which is difficult to detect. On the other hand, early detection of MIC problems is imperative, due to the fact that a complete eradication of bacteria, once they are well established in a system, is nearly impossible. Biocides are selected on the basis of the following criteria:

1 Effectiveness;
2 Economics;
3 Safety;
4 Compatibility with system fluids; and
5 Compatibility with other treatment chemicals.

In Tables 3–6 the most frequently used biocides are listed, and a short summary of advantages and disadvantages is given [9].

Table 3. Oxidising biocides [3]

Biocide	Advantages	Disadvantages
Chlorine	– Cheap and effective – Broad-spectrum activity – Simple monitoring	– Operator hazards – Ineffective against bacterial biofilms
Chlorinating compounds	– Less hazardous handling – As effective as chlorine	– Scales problems – Expensive – Larger quantity needed than Chlorine
Chlorine Dioxide	– pH insensitive, dissolves FeS – Good against biomasses – Compatible with organic compounds	– Toxic – Expensive – Difficult to handle
Chloramines	– Good against biofilm – Good persistence – Reduced corrosivity – Low toxicity	– Ammonia injection is required – More expensive than chlorine
Bromine	– More effective than Cl_2 at higher pH – Broad-spectrum activity	– Same disadvantages than Cl_2 – Expensive

Water chlorination is the simplest and more widely used biocide. For stainless steels, however, there is a risk of increasing corrosion susceptibility as chlorine concentration increases. Chlorine is added to sea water at ppm level. The consumption of the product is rapid, due to the reactions of the halogen with organic and inorganic species present in the sea water, mainly of three kinds:

(a) Bromides;
(b) Ammonia; and
(c) Organic substances.

The decay of the oxidant can be quick, in the order of 10–20 min [10]. To keep the amount of chlorine as low as possible, but at the same time effective, it is important to be able to measure the residual oxidant concentration.

4. Antifouling Paints

Antifouling paints are used for ships and marine structures. They act by releasing bioactive materials which are bonded to the structure dispersed in a matrix. The release mechanisms are responsible of the extent of drydocking intervals. Economical and environmental issues are the most important issues in the selection of biocide paints.

Table 4. Non-oxidising biocides [3]

Biocide	Advantages	Disadvantages
Aldehydes		
Formaldehyde	– Cheap	– Suspected carcinogen – High dosages required – Reacts with NH_3, H_2S, oxygen scavenger

Glutaraldehyde	– Broad spectrum activity
	– Insensitive to Sulfides
	– Nonionic
	– Tolerates salts and hardness
		– Deactivated by:
		  * Ammonia
		  * Primary amines
		  * Oxygen scavengers

Acroleine	– Broad spectrum activity – Dissolves sulfides, penetrates deposits	– Highly toxic – Reactive – Difficult to handle
Amine-type compounds		
Amine and diamine	– Broad-spectrum activity – Inhibition properties – Effective in water containing sulfides – Surface active	– React with anionic chemicals – Less effective in presence of suspended solids
Quaternary amine compounds	– Broad-spectrum activity – Good surfactancy – Good persistence – Low reactivity with other chemicals	– Inactivated in brines – Foaming – Slow acting

Bioactive materials are usually metallic and/or organometallic — in particular, compounds of copper, arsenic, mercury, zinc, antimony and lead. An example of chemical biocides used in paints include cuprous oxide and triorganotin. They are highly toxic and extremely polluting. Alternative chemicals are under investigation with the aim of achieving a comparable efficiency with a lower toxicity [11].

Four different type of antifouling paints are mainly used [8]:

(i)   Conventional, soluble matrix antifouling: in those paints, the biocide is released together with the soluble binder of the matrix; the lifetime is about one year.

(ii) Advanced, insoluble matrix antifouling: in these products the binder is no longer soluble in sea water; biocides and other soluble ingredients dissolve from the surface layer; the useful lifetime is about two years.

(iii) Self-polishing antifouling: these products are based on a matrix, e.g. organotin polymer, of remarkable mechanical strength, which allows the build-up of a very thick layer. The time of release is about 5 years.

(iv) Polishing/ablative antifouling: these products are based around the same principles as self-polishing antifouling and works with a polishing/ablation mechanism, offering the best performance without the use of organo–tin components, which are highly toxic and restricted by many regulatory laws[12].

5. Environmentally-conscious Alternatives for Biocide Actions

For a lower impact on the environment there is a strong demand for further research. New biocides need to be developed, having the following characteristics:

Table 5. Non-oxidising biocides [9]

Biocide	Advantages	Disadvantages
Halogenated compounds		
Bronopol	– Broad spectrum activity – Low toxicity – Ability to degrade	– Dry chemical – Breaks down in high pH
DBNPA	– Broad spectrum activity – Fast active and effective	– Expensive – Affected by sulfides
Sulfur compounds		
Carbamates	– Active against SRB – Effective in alkaline pH – Useful for polymer solutions	– High concentration required – Reactive with metal ions
Isothiazolone	– Broad spectrum activity – Compatible with brines – Good against sessile bacteria – Low dosages, degradable	– Expensive – Cannot be used in sour systems
Metrodinazole	– Active against SRB – Compatible with other chemicals	– Specific to aerobic organisms

Table 6. Non-oxidising biocides. Quaternary phosphonium salts (quite new – used in the injection water)[9]

Biocide	Advantages	Disadvantages
Quaternary phosphonium salts	– Broad spectrum activity – Good stability – Low toxicity – Not affected by sulfides	– Not yet known

1   Efficacy;
2   Economy;
3   Low mammalian toxicity;
4   Low impact;
5   Inactivation in the environment.

Two possible directions for the research seem to be promising: the use of micro (or macro) organisms against fouling (such as the biological pest control in agriculture) and the development of bioactive materials from marine organisms.

In the case of marine environments, it may be interesting to find out how some marine organisms manage to stay free from microbial colonisation in an open non-sterile system, where practically every surface is covered by microorganisms. For instance, some researchers [13] have investigated the fact that certain gorgonian octocorals are naturally fouling-resistant. They state that biofouling is a complex phenomenon and that is improbable that only one strategy could effectively inhibit the wide range of potential fouling organisms. They suggest that "minimally adhesive" surfaces based on low energy outermost layers may be a primary, passive antifouling defence in gorgonian corals. Other marine organisms protect their surfaces in a number of ways:

(i)   by the shedding of their skin;

(ii) by the excretion of antimicrobial, antialgal and antimicotic agents;

(iii) by the excretion of proteolytic enzymes, which block the movement of cilia colonising larvae, acting as a "polish"and protecting against small crustaceae and mites, which graze on algae and fungi and break through the defense mechanisms.

Research has been recently carried out (in the framework of an EC funded research project) in Eniricerche with the aim of assessing the effectiveness of the chemical strategy that the algae seem to possess for discouraging the settlement of a microbial biofilm on their surfaces [14].

Different raw fractions obtained by processing the marine benthic diatom *Amphora coffeaeformis* were screened for anti-slime activity against marine biofilms grown in laboratory-simulated sea water environments.

The conclusions of the first phase of the project indicate that algal cell-free cultural waters showed a detrimental action on the microbial adhesion rates. This effect was particularly evident in the period between the 20th and 40th day of immersion time.

The antifouling effect of the algal waters was confirmed by antimicrobial testing, by microbial direct counts and by SEM observations.

According to the literature, the chemical family of the active substances produced by the *Amphora* cells has been classified as phenolic derivative.

6. Conclusions

The possible options for the selection of biocides in the oil industry in relation to oilfield problems and to offshore components are very large. Some of them, with selection criteria, are presented in this paper.

In most cases it is very difficult to prevent the accumulation of biofilms. So bearing in mind that "The organism always wins" (Kevin C. Marshall's comment on the prospects of biofouling control), it is important to design a strategy to keep the accumulation of biofilm below a level of interference with the production demands.The research in this field is far from complete; in particular, there is a need for:

(a) A better understanding of active mechanisms, so that the biocides can be used in small quantities and with the maximum of efficacy;

(b) A more conscious attitude toward environmental issues, through the evaluation of existing substances and the study of natural biocides, such as those produced by marine organisms.

The goal will be, learning from natural systems, to know how to live together with biofouling.

5. Acknowledgements

The authors thank AGIP S.p.A (I. Obracaj, C. Di Iorio) for the useful information and discussions.

References

1. W. G. Characklis and K. C. Marshall, Biofilm: a basis for an interdisciplinary approach, in *Biofilms* (eds W. G. Characklis and K. C. Marshall), John Wiley & Sons, Inc., 1990.
2. V. Scotto *et al.*, Microbial and biochemical factors affecting the corrosion behaviour of stainless steels in sea water, in European Federation of Corrosion Publication No. 10, 21–35. The Institute of Materials, London, 1993.
3. NACE TPC 3 Publication, Microbiologically Influenced Corrosion and Biofouling in Oilfield Equipment, 1990, NACE, Houston, Tx.
4. E. B. Kajer, *Progress in Organic Coatings*, 1992, **20**, 339–352.
5. V. Rascio, C. Giudice and B. Del Amo, *Progress in Organic Coatings*, 1990, **18**, 389–398.
6. H. C. Flemming and G. G. Geesey, Biofouling and biocorrosion in industrial water systems, in *Proc. Int. Workshop on Industrial Biofouling and Biocorrosion*, Stuttgart, 13–14 September, 1990. Published by Springer-Verlag.
7. R. Johnsen, North sea experience with the use of stainless steels in sea water applications, in European Federation of Corrosion Publication No. 10, 48–59. Published by The Institute of Materials, London, 1993.

8. P. Gallagher, A. Nieuwhof and R. J. M. Tausk, Experiences with sea water chlorination on copper alloys and stainless steels, in European Federation of Corrosion Publication No. 10, 73–91. The Institute of Materials, London, 1993.

9. J. Boivin, Oil Industry Biocides, *Mater. Perform.*, Feb. 1995, 65–68.

10. C. Madec, F. Quentel, and R. Riso, Sea water chlorination, in European Federation of Corrosion Publication No. 10, 60–72. The Institute of Materials, London, 1993.

11. E. Lindner, The role of surface free energy in development of non-toxic antifoulants, *8th Int. Congr. on Marine Corrosion and Fouling*, 21–25 September 1992, Taranto, Italy. Published by Consiglio Nazionale delle Richerche – Instituto Sperimentale Talassografico.

12. J. A. Lewis and I. J. Baran, Biocide release from antifouling coatings, *8th Int. Congr. on Marine Corrosion and Fouling*, 21–25 September 1992, Taranto, Italy.

13. N. H. Vrolijk, N. M. Target, R. E. Baier and A. E. Meyer, Surface characterization of two Gorgonian coral species: implication for a natural antifouling defence, *Biofouling*, 1990, **2**, 39–54.

14. B. Pietrangeli, M. Camilli and R. Gianna, Antifouling action of algal substances, *7th Europ. Congr. on Biotechnology*, Nice, France, 19–23 February 1995. Published in the ECB (European Conferences on Biotechnologies) Series.

5

Practical Consequences of the Biofilm in Natural Sea Water and of Chlorination on the Corrosion Behaviour of Stainless Steels

T. ROGNE and U. STEINSMO

SINTEF Corrosion and Surface Technology, N-7034 Trondheim, Norway

ABSTRACT

More than one decade of intensive research at SINTEF Corrosion and Surface Technology on the local corrosion susceptibility of stainless steels exposed to natural and chlorinated sea water is reviewed. In natural sea water it is shown that the formation of a biofilm on the surface has an important influence on the local corrosion initiation tendency and propagation rate for temperatures up to about 30°C. In chlorinated sea water the tendency towards local corrosion initiation is even stronger, but the propagation rate is typically less than in natural sea water, and more dependent on the cathode/anode area ratio. Surface preparation and ageing of the surface is shown to have a distinct influence on the crevice initiation tendency, and this observation may be utilised during the start-up of a chlorinated sea water handling system in order to minimise the risk of local corrosion. In chlorinated sea water systems, corrosion can be avoided and the application area of stainless steels can be extended by use of the Resistor-controlled Cathodic Protection (RCP) method, developed at SINTEF Corrosion and Surface Technology.

1. Introduction

Testing of stainless steel for sea water applications may involve different types of tests, ranging from long term immersion tests in natural sea water to short time tests in more artificial electrolytes. When properly used, most of the tests will reveal significant information about the corrosion resistance of the steels in sea water.

The most important type of corrosion on stainless steels in sea water is local corrosion, in particular crevice corrosion and pitting at welds. Over the last 15 years much work has been carried out at SINTEF Corrosion and Surface Technology related to these types of corrosion attacks. Some of the investigations have been rather basic in nature, focusing on the electrochemical behaviour of an oxidised stainless steel surface in various environments — such as natural sea water, chlorinated sea water and simulated crevice solutions. Other studies have been more practical, trying to establish a better knowledge on how to avoid local corrosion in real systems [1–14].

The present paper emphasises some aspects and consequences of both biofilm formation and chlorination on the corrosion properties of the stainless steels for different sea water applications. The paper is divided into three parts — in the first part, results of tests in natural sea water are discussed, with emphasis on the cathodic effect of a biofilm. The second part focuses on the behaviour of stainless steels exposed in chlorinated sea water, which is recognised to be an even more corrosive medium than natural sea water. In the third section, the application of cathodic protection to chlorinated sea water systems is discussed, where protection can be easily achieved by the use of Resistor-controlled Cathodic Protection (RCP).

2. Exposure Tests in Natural Sea Water

When a stainless steel surface is exposed to natural sea water at moderate temperatures the corrosion potential is observed to rise with time, as shown in Fig. 1. At temperatures below 32°C the potential rises to about 300 mV SCE within 3–12 d, depending upon the temperature. Prolonged exposure has shown that the potential may continue to rise slowly to near 400–430 mV SCE after more than one year [2]. This potential rise has been observed in many different laboratories [15–20 ,23], and has in all cases been attributed to the formation of a biofilm on the surface.

   The marked difference in the potential level at 32°C compared to lower temperatures indicates that the biofilm effect is absent at 32°C. This is further supported by the results in Fig. 2, from tests performed in substitute sea water, where no sea water biomaterial is present. In this Figure, there is a small potential rise and a temperature effect, though less evident at low temperatures as shown in Fig. 1. The results of Fig. 2 can be related to other observations which show that the passive current density increases with temperature [10] and decreases with time [5].

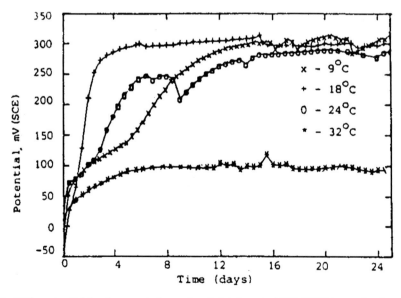

Fig. 1 *The potential developement of samples of stainless steel 254 SMO in nearly stagnant sea water, at four different temperatures [5].*

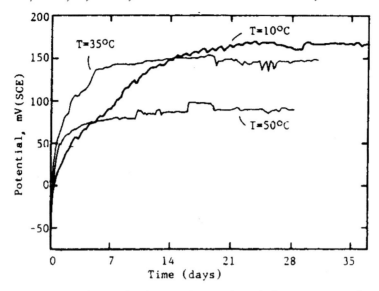

Fig. 2 Potential vs time for samples of 254 SMO exposed to substitute sea water, at three different temperatures: 10, 35 and 50°C [5].

The potential rise in the presence of a biofilm has been observed on many different qualities of stainless steels, nickel base alloys and also on titanium [3, 5]. This generality indicates that the effect is connected with changes in the cathodic reaction rather than the anodic reaction. This implies that most materials (in the passive state) will experience a similar potential rise. Copper alloys are an exception here [3], possibly due to the poisoning effect of the copper ions.

The practical consequence of the bioactivity is that passive materials at moderate temperatures will be subject to an increased risk of local corrosion initiation due to the rise in potential. High-alloyed stainless steels, such as 6 Mo and superduplex, do not normally corrode in natural sea water at temperatures up to 30°C, but the more conventional stainless steels (AISI 304 or 316) are not resistant enough to withstand a potential rise up to 300–430 mV SCE, even in rather cold sea water.

The existence of a maximum temperature for an "active" biofilm has also been observed also by Scotto *et al.* [21]. However, the maximum temperature in the Mediterranean water seem to be somewhat higher than 32°C — most probably attributed to the higher tolerance limit of the bacteria in the generally warmer sea water on the Italian coast. The results by Scotto *et al.* further indicate that the biofilm effect has a maximum limit of about 2.5 ms^{-1} with respect to the sea water flow rate. Our investigations have revealed no such limit in the range of flow rates up to 3.8 ms^{-1} [5].

The interpretation of the biofilm effect as a result of a modification of the cathodic reaction has not been based on the measurements of the potential rise alone. Cathodic polarisation measurements on plate samples, both potentiodynamic and potentiostatic, are in support of this. Figure 3 shows some typical results obtained with relatively slow potentiodynamic polarisation in natural sea water. The curve

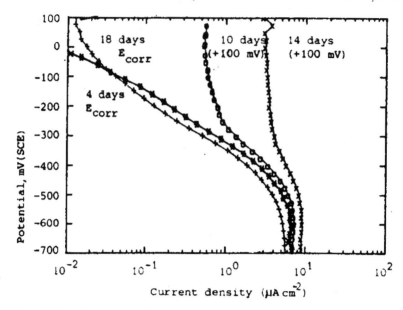

Fig. 3 Potentiodynamic curves for 254 SMO steel recorded after 1 and 18 d of free exposure and at 10 and 14 d at +100 mV SCE respectively. Nearly stagnant sea water and temeprature 18±1 °C [5].

marked (4 d, E_{corr}) is a typical polarisation curve in the absence of a biofilm, as similar curves are obtained in substitute sea water. The curve is characterised by a Tafel type behaviour down to about –500 mV SCE, below which there is a limiting oxygen current density. After 18 d exposure, i.e. when the potential has increased and levelled out, there is only a small change in the polarisation curve at the smallest current densities (highest potentials > –100 mV SCE), where the curve bends upwards. This extra contribution can explain the potential rise since the corrosion potential is a mixed potential due to the cathodic reaction and an anodic reaction, which here is a passive current density in the order of 0.01–0.001 µA cm^{-2} [5].

What is particularly interesting in the results of Fig. 3 is the more dramatic changes of the cathodic properties following a cathodic polarisation to 100 mV SCE. The previous small reaction rate at the upper potentials increases by several decades, and after a couple of weeks the whole cathodic curve is almost vertical and little dependent on the potential. The effect is also sensitive to the temperature in the same way as the potential rise as shown in Fig. 4. This strongly supports the idea that the potential rise and the cathodic reaction increase at the upper potentials are both due to the same biofilm-related mechanism. The details of the mechanism are not clearly understood, but some ideas have been discussed elsewhere [4].

From a practical point of view, the cathodic reaction enhancement has serious implications. If local corrosion occurs the potential will normally drop to a lower level until the available cathodic current balances the anodic current of the local pit or crevice. A large cathodic efficiency, as observed in the presence of a biofilm, means that a rather low area ratio between a cathode and an active anode is sufficient to

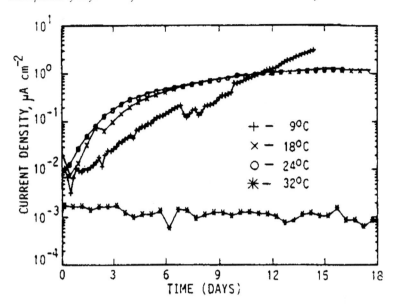

Fig. 4 The development of the cathodic CD on stainless steel 254 SMO at +100 mV SCE (0 mV at 32 °C). Pre-exposed for 14 d of free corrosion potential. At nearly stagnant conditions and 4 different temperatures [5].

allow rather large corrosion rates. This is demonstrated in Fig. 5, where actively corroding crevice samples of AISI 316 were coupled to an outer cathode. With area ratios as low as 5–100 the corroding crevices in natural sea water experienced an increase in the corrosion current up to 100–300 μA cm^{-2} or 1–3 mm/y after the initial 7–10 d time lag when the biofilm was forming. In contrast, the crevice exposed in substitute sea water (ASTM D1141) remained at a constant corrosion rate about two decades lower, even with the much larger area ratio of 2000:1.

A similar development of the corrosion rate will be seen when a small 90/10 CuNi anode or other materials that corrode in the potential range from –300 to +100 mV SCE, are coupled to a larger stainless steel cathode in sea water.

The initiation of crevice corrosion is to a great extent dependent on the incubation time before the ennoblement starts and how fast the potential is increased. This is clearly shown in Fig. 6. Crevice corrosion was not observed in the spring test, while in the summer test, where the ennoblement is faster, all 316 L samples initiated crevice corrosion [22].

The biofilm formed on the metal surface in natural sea water is due to the settlement of bacteria [6,23]. Thus, the biofilm effect should be considered as an example of MIC (Microbial Influenced Corrosion). In the literature MIC is a subject of great concern, and a number of methods are used for its characterisation, as reviewed by Mansfeld and Little [24]. In natural sea water there seems to be little doubt that the main effect of the bacterial activity is on the enhancement of the cathodic reaction. It is therefore surprising that measurements of cathodic reaction rates have not been utilised to a greater extent. In our opinion it is one of the most sensitive indicators of

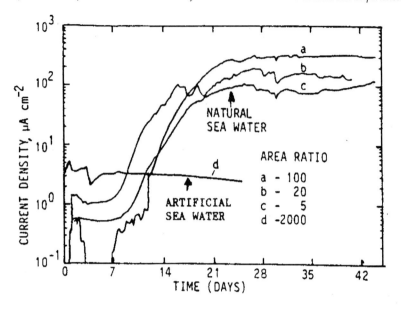

Fig. 5 *Crevice corrosion rate of AISI 316L stainless steel in nearly stagnant natural and substitute sea water at different ratios between cathode and anode area. Temperature 8±1 °C [5].*

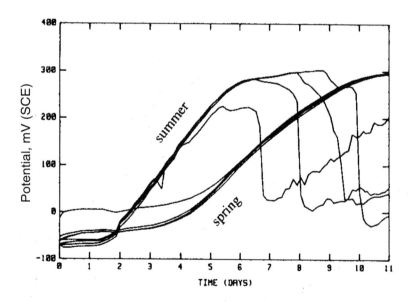

Fig. 6 *Exposure of AISI 316 L in natural sea water. Crevice corrosion was initiated during the summer test but not during the spring test [22].*

the bio-activity that relates directly to the local corrosion susceptibility.

3. Exposure Tests in Chlorinated Sea Water

Chlorine is added to sea water to kill the living organisms, including bacteria. In this way the biofilm effects discussed above will not occur. However, the chlorine causes a series of new cathodic reactions, some of which will have a very high redox potential. The result is that the passive or non-corroding stainless steel is raised to even higher potentials than in natural sea water. The rise in potential is dependent on the material properties, but in general no major difference has been observed for the highly alloyed stainless steels. With respect to environmental parameters, the actual potential level is a function of the chlorine concentration and the temperature, as shown in Fig. 7. These tests were conducted in slowly moving, nearly stagnant sea water. A typical concentration for practical use is 0.5–1.0 ppm, and typical operating temperatures are 20–30°C. According to Fig. 7, we should then expect potentials in the range 600–650 mV SCE. However, the potential is also affected by the flow rate. Later research [25] has shown that an increase in the flow rate from 0.05–0.1 ms^{-1} to 3.2 ms^{-1} leads to a decrease of the potential in the range 30–50 mV. Due to the high potentials, it is no surprise that local corrosion is initiated more easily in chlorinated sea water than

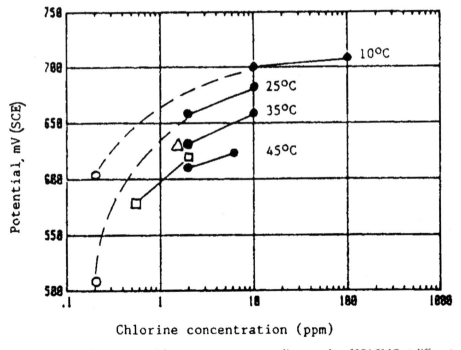

Fig. 7 Near steady state potentials measures on non-corroding samples of 254 SMO at different levels of chlorination and temperature. • Gartland and Drugli [7], 14 d; o R. Gundersen et al. [6], 50 d; ☐ Wallen and Henriksen [28], 45°C, 90 d; △ Wallen and Henriksen [28], 35°C, 90 d.

in natural sea water. The conventional stainless steels such as AISI 304 and AISI 316 have no chance to resist crevice corrosion initiation even in cold sea water. The more highly alloyed stainless steels like the 6-Mo stainless steels and the so-called superduplex stainless steels can still be used, but at a somewhat lower temperature than in natural sea water. Norsok [26] recommends 15°C as the upper temperature limit for 6 Mo and super duplex stainless steel (threaded connections) for application in chlorinated sea water.

The results of Fig. 7 were in part obtained in a series of experiments designed to compare the corrosion resistance of SAF 2507 and 254 SMO. The experimental arrangement is illustrated in Fig. 8. Crevice samples of the actual materials were immersed in heated, chlorinated sea water and coupled to large cathode plates of the same material. The potential of the whole assembly as well as the current flowing between the crevice sample and the cathode were monitored. Initiation of corrosion was observed from the current readings without having to remove the samples from the sea water. The test series was run in six different conditions, each of two weeks duration. The six conditions were nominally 25, 35 and 45°C, with two chlorine levels 2 and 10 ppm at each temperature. Two types of samples were used. The "old samples" were never taken out of the sea water but were exposed to all six conditions of gradually increasing severity. The "new" samples were used for only one condition, i.e. for each new condition these samples had never been exposed before. The testing was carried out with three parallel samples at each condition. Further details are given elsewhere [7]. The results were not particularly surprising in respect of the behaviour of the "new" samples of both materials. Figure 9 shows that the new samples of 254 SMO resisted 35°C and 2 ppm at the most, and SAF 2507 was slightly better. The most surprising observation was, however, that all parallel samples of the old samples of both materials resisted the most severe conditions. This observation is attributed to an ageing effect of the oxide which then became more resistant on prolonged exposure in chlorinated sea water. The mechanism of this ageing was not studied here, but it is reasonable to relate the effect to changes of the oxide thickness and composition caused by the repeated potential rise.

The consequence of this observation for testing is that fresh samples should always be used when changing the testing conditions in order to obtain the lower limit of the acceptable environmental conditions. On the other hand, a practical consequence of the ageing effect is that the most critical phase for stainless steel materials in a chlorination plant will be in the first few weeks of operation. If crevice corrosion attack can be avoided at the onset, the chances of getting corrosion attack at a later stage will be reduced, provided that the environmental parameters are not altered towards more severe conditions.

The crevice geometry using rubber bound aramid as gasket material has been found to give a rather mild crevice condition, leading to high critical temperatures for onset of corrosion. Compared with other gaskets, such as POM or Teflon, the critical crevice temperatures are decreased [36]. Graphite type gaskets are even worse [36], as shown by the results in Fig. 10. These data clearly show the importance of applying the correct gasket material with respect to corrosion severity. In the NORSOK standards graphite gaskets are not recommended for use.

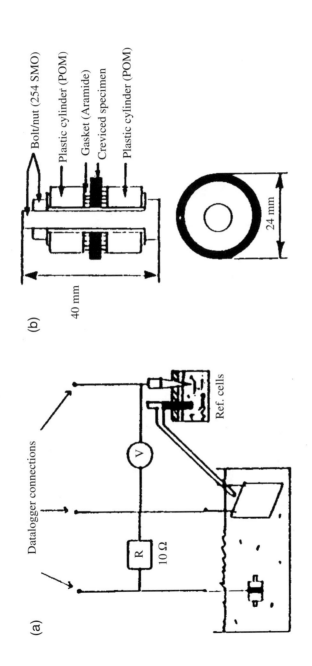

(a)

Datalogger connections

(b)

Bolt/nut (254 SMO)
Plastic cylinder (POM)
Gasket (Aramide)
Creviced specimen
Plastic cylinder (POM)

40 mm

24 mm

R
10 Ω

V

Ref. cells

Fig. 8 Schematics of the sample mounting and the measuring system for remote crevice assembly testing in sea water [7].

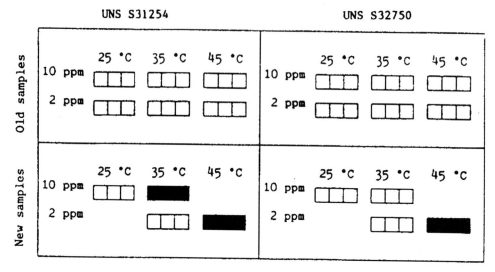

Fig. 9 *A summary of test results with crevice samples in chlorinated sea water. An open square means a non-corroding sample while a filled square means a corroding sample [7].*

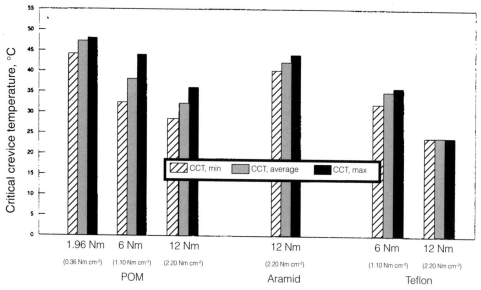

Fig. 10. *Effect of torque and gasket material on the crevice corrosion temperature of a 6 Mo stainless steel appling the new SINTEF test method [36].*

Up to now we have referred only to the potential level when discussing the local corrosion susceptibility. In chlorinated sea water the potential rise takes place much faster than in natural sea water. As shown in Fig. 11(a) the potential rise in chlorinated sea water takes place within a few hours, as compared to a few days in natural sea

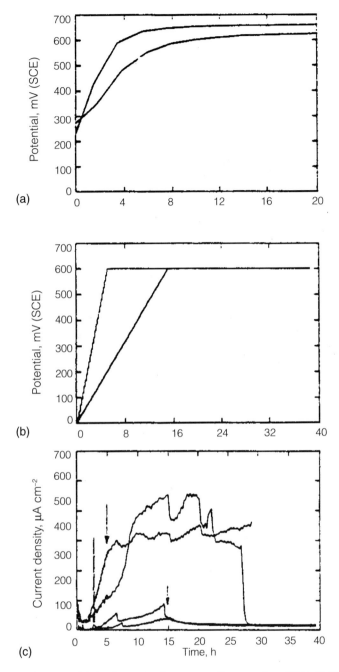

Fig. 11 (a) The potential development of non-corroding samples of 254 SMO in sea water at 35 °C with 2 and 10 ppm chlorine [8]. (b) Fast and slow potential rise from 0 to 600 mV SCE simulated with a computer controlled potentiostat. (c) Anodic current densities for two of three parallel crevice samples of 250 SMO exposed to a 3% NaCl solution at 35 °C. The three samples in each test series were subject to the two different potential developments as given in (b) [8].

water. This rapid potential rise has a direct influence on the local corrosion initiation tendency, at least for crevices. The reason is that the rate of potential rise, dE/dt, adds an extra contribution to the passive current density [8]. With a larger passive current density there is an increased probability of getting crevice corrosion initiation. This is demonstrated in Fig. 11. The results of Fig. 11 are taken from an experiment in which three parallel specimens were exposed to a NaCl solution at 35°C with no chlorine present [8]. The potential rise with 2 ppm and the more rapid potential rise with 10 ppm chlorine were simulated using a potentiostat and a computer. As shown in Fig. 11(c), two out of three specimens that were subject to the fast potential scan initiated local corrosion — not observed in the specimens subject to the slow potential scan. It should be noted that the final potential level was the same in both cases (600 mV SCE), simulating the conditions in chlorinated sea water.

The experiment shows that corrosion initiation of stainless steels has the highest probability during the initial potential rise period, and during variation in temperature, chlorine concentration, etc. This is in agreement with other observations that initiation normally takes place within rather short times in chlorinated sea water (1–4 d), if it ever occurs [27, 28]. For good reproducibility and reliability of the test results one should design the experiment in such a way that there is no unintended reduction of the test severity. This would be the case if the chlorine were added to cold water which was later heated to the test temperature over a time longer than the time of the potential rise, or, if the water were heated, and then the chlorine concentration was raised gradually over a period longer than the typical time of the potential rise. It may be pointed out that such a smooth adjustment of the conditions could be utilised as a positive remedy to reduce the risk of corrosion during the start-up phase in real systems carrying chlorinated sea water.

It should also be noted in the context of this discussion that the potential rise is faster for "old" systems, due to the fact that the anodic passive current is reduced with time. In service the procedures for fire water systems are often such that fresh chlorinated sea water is added at intervals from a few days to a few weeks. During the intervals the chlorine and oxygen are consumed in the cathodic process, leading to a decrease in the potential. Each time fresh chlorinated sea water is added, the potential will therefore rise very rapidly, with a high risk — dependent on the temperature for initiation of crevice corrosion. This is probably the main reason for many of the corrosion failures experienced in the North Sea in fire water systems.

With respect to crevice corrosion propagation in chlorinated sea water, the cathodic efficiency is of equal importance as in natural sea water. Figure 12 shows some cathodic polarisation data obtained in flowing chlorinated sea water [25]. It can be seen that the chlorine has an effect over the whole potential region. In the propagation phase of crevice corrosion the potential will be lowered to somewhere in the region of −200 to +200 mV SCE. In this region the cathodic efficiency in chlorinated sea water is much less than in natural sea water with an active biofilm as seen in Fig. 12. This implies that, as long as the corrosion rate is controlled by the cathode, the corrosion rates will be much lower in chlorinated sea water than in natural sea water up to 32°C. This is clearly verified by the results in Fig. 13, where active crevice assemblies of AISI 316 were coupled to remote cathodes with a fixed area ratio. In

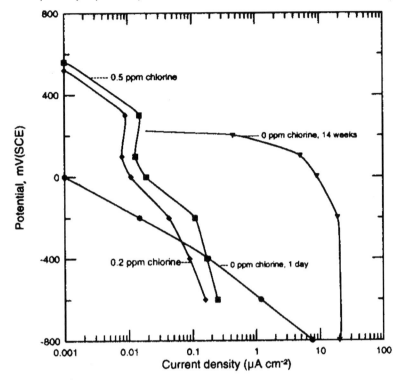

Fig. 12 *Polarisation data from potentiostatic tests in chlorinated sea water 30°C for free residual chlorine concentrations of 0.2 and 0.5 ppm and exposure times of 47 and 54 d respectively. Data obtained in natural sea water at 9°C for exposure times 1 d and 14 weeks are also shown [25].*

the same way as shown in Fig. 5, the corrosion rate in natural sea water increases to very high levels, up to 3–3.5 mm/y for a cathode/anode area ratio of 182, while it is much lower in chlorinated sea water. The results are summarised in Fig. 14 for different anode/cathode area ratios.

In a real chlorination system comprising extended lengths of pipes, tanks, heat exchangers etc. the area ratio between a corroding spot and the cathode can be quite large. Simulation calculations have shown that a cathode/anode area ratio of 10.000:1 or even larger is necessary to achieve maximum corrosion rates in chlorinated sea water pipe systems [11,12] and as indicated by the test results shown in Fig. 14. For pipe systems the effective cathode area will be dependent on the pipe diameter and conductivity of the sea water, due to a potential drop. This is shown in Fig. 15(a). On the other hand, in natural sea water maximum corrosion rates were achieved with an area ratio much less than that in chlorinated sea water as shown in Fig. 15(b). This difference in behaviour should be borne in mind when designing propagation tests and analysing propagation test results. With the relatively small area ratios used in testing (typically 100:1 or lower) corrosion rates measured in chlorinated sea water will tend to be much lower than in a real system. However, in natural sea water up

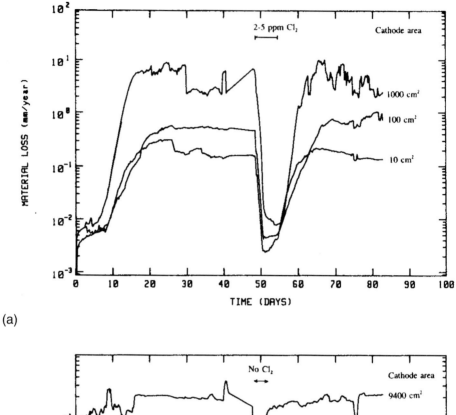

(a)

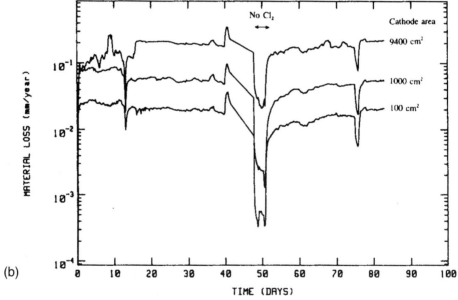

(b)

Fig. 13 Corrosion rate for crevice specimens of AISI 316L stainless steel in natural sea water; (a) interrupted by a chlorination period, and in sea water with 0.5 ppm residual chlorine, and (b) interrupted by a period without chlorine [36].

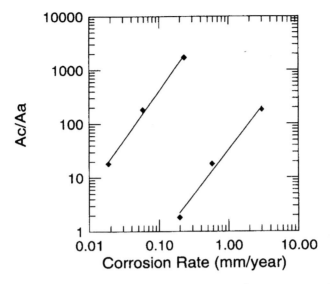

Fig. 14 *Corrosion rate of AISI 316 L in slowly flowing natural (left) and chlorinated sea water at 10–15°C at different cathode/anode area ratios [36].*

to about 32°C the test results will be more comparable with those for a large plant.

4. Internal Cathodic Protection of Pipelines

Initiation of local corrosion of stainless steel can be prevented by applying RCP – Resistor-controlled Cathodic Protection. The RCP method has been developed essentially as a mean to prevent local corrosion on stainless steels and galvanic corrosion in piping systems with various types of saline water, in which a critical combination of the potential and the temperature may be exceeded [30–34].

The RCP method is based on a patent claim from SINTEF: "Method and arrangement to hinder local corrosion and galvanic corrosion in connection with stainless steels and other passive materials" [30]. The method is now being industrialised in cooperation between SINTEF Corrosion and Surface Technology and CorrOcean A/S. The first RCP systems have already been taken into service [35].

The basic principle of the method is to apply cathodic protection to a pipe system of stainless steel using a resistor in series with the anode to control both the potential near the anode and the anode current output. The principle is shown schematically in Fig. 16 [30].

The method is based on the observation that the protection potential for prevention of local corrosion on stainless steel is much more positive than the typical potentials of sacrificial anodes. The voltage drop over the resistor is therefore designed to obtain sufficient, but not excessively negative, polarisation of the stainless steel. The resistor

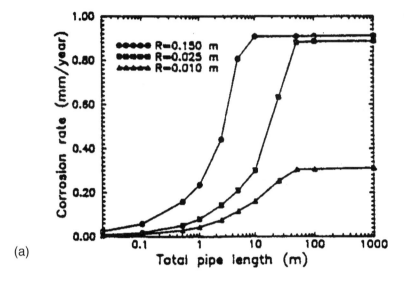

(a)

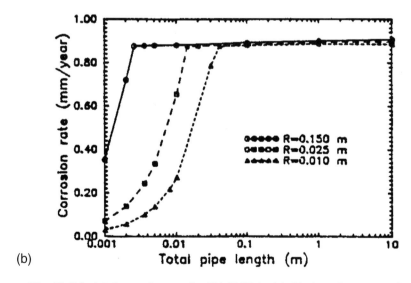

(b)

Fig. 15 *Calculated corrosion rate for 254 SMO in (a) chlorinated sea water (1 ppm residual chlorine), and (b) natural sea water at 25 °C, for different pipe geometries [11].*

control thus keeps the stainless steel in a protective potential range, where the current requirements are very small in many saline environments, as in chlorinated sea water, produced water from the oil and gas production and natural sea water above about

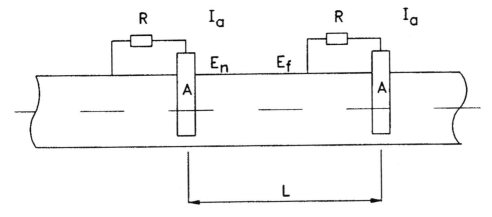

Fig. 16 *Schematic of the RCP method applied to a pipe system [30].*

30°C. For chlorinated sea water this is shown by the test data in Fig. 12. Due to the very low current requirements in the relevant potential range, a single anode can protect large lengths of the pipe systems at a very low anode consumption rate.

Another advantage of the application of RCP is that lower alloyed stainless steel can be used as an alternative to higher alloyed and more expensive materials for many applications, which thus gives a large economic perspective. The application area for stainless steel will be extended for instance toward higher temperatures, higher chlorination levels and will therefore resist larger variations in the service conditions, giving a more reliable and safe service.

6. Conclusions

• For temperatures up to about 30°C an "active" biofilm is formed on stainless steel surfaces exposed to natural sea water. This biofilm stimulates the cathodic reaction. The potential rise to 300–350 mV SCE, that takes place within a few days, increases the risk for onset of local corrosion. The enhancement of the cathodic reaction rate results in anodically controlled propagation rates for very low cathodic areas.

• As well as the ennoblement of the potential, the cathodic reaction rate on stainless steels polarised from –100 to 100 mV SCE in natural sea water, is a very sensitive indicator of an "active" biofilm on the surface.

• In chlorinated sea water the potential of non-corroding stainless steels rises to the 500–650 mV SCE region within a few hours, dependent on the conditions, and further increases the risk of local corrosion initiation as compared with natural sea water. However, the cathodic reaction rate in chlorinated sea water is much less than in natural sea water up to about 30°C. The corrosion propagation rate in chlorinated sea water will therefore be under cathodic control unless the cathodic area is very large.

• The potential rise period is very important for the susceptibility to local corrosion initiation.

• The crevice corrosion initiation is very sensitive to the crevice geometry and is thereby affected by the gasket material and torque.

• Stainless steels become more resistant to local corrosion initiation with prolonged exposure under non-corroding conditions.

• Initiation of local corrosion of stainless steels exposed to chlorinated sea water can be prevented by applying RCP — Resistor-controlled Cathodic Protection.

• With application of RCP, lower alloyed stainless steel can provide an alternative to higher alloyed and more expensive materials for many applications, providing a large economic perspective. The application area for stainless steel will be extended, for instance, toward higher temperatures, higher chlorination levels and will therefore resist larger variations in the service conditions, giving a more reliable and safe service.

References

1. R. Johnsen, E. Bardal and J. M. Drugli, Cathodic Properties of Stainless Steel in Sea Water, in *Proc. 9th Scand. Corrosion Congr.,* Copenhagen, September, 1983.
2. R. Johnsen and E. Bardal, Cathodic Properties of Different Stainless Steels in Natural Sea water, *Corrosion,* 41, 5, 1985.
3. R. Holthe, E. Bardal and P. O. Gartland, Time Dependence of Cathodic Properties of Materials in Sea water, *Mater. Perform.,* 1989, **28**, 6, 16.
4. R. Holte, P. O. Gartland and E. Bardal, Oxygen Reduction on Passive Metals and Alloys in Sea water — Effect of Biofilm, in *7th Int. Congr. on Marine Corrosion and Fouling,* Valencia, Spain, November 1988.
5. R. Holte, The Cathodic and Anodic Properties of Stainless Steels in Sea Water; dr.ing. thesis 1988, Dept of Materials and Processes at the Norwegian Institute of Technology, University of Trondheim, Norway.
6. R. Gundersen, B. Johansen, P. O.Gartland, I. Vintermyr, R. Tunold and G. Hagen, The Effect of Sodium Hypochlorite on Bacterial Activity and the Electrochemical Properties of Stainless Steels in Sea water, *Corrosion '89,* Paper 108, NACE, Houston, Tx, 1989.
7. P. O. Gartland and J. M. Drugli, Crevice corrosion of high-alloyed stainless steels in chlorinated sea water. Part I — Practical aspects, *Corrosion '91,* Paper 510, NACE, Houston, Tx, 1991.
8. P. O.Gartland and S. I.Valen, Crevice corrosion of high-alloyed stainless steels in chlorinated sea water, Part II — aspects of the mechanism, *Corrosion '91,* Paper 511, NACE, Houston, Tx, 1991.
9. P. O. Gartland, R. Holte and E. Bardal, Evaluating Crevice Corrosion Susceptibility from Testing in Simulated Crevice Electrolytes and Mathematical Modelling, *11th Scand. Corrosion Congr.,* Stavanger, 1989.
10. S. I. Valen and P. O. Gartland, Critical Temperatures for Crevice Corrosion of High Alloyed Stainless Steels in Sea Water, *EUROCORR'91,* Budapest, 1991.

11. S. I. Valen, Initiation, Propagation and Repassivation of Crevice Corrosion of High-Alloyed Stainless Steels in Sea water; dr.ing. thesis 1991, Dept of Materials and Processes at the Norwegian Institute of Technology, University of Trondheim, Norway.

12. P. O. Gartland and J. M. Drugli, Methods for Evaluation and Prevention of Local and Galvanic Corrosion in Chlorinated Sea water Pipelines, *Corrosion '92*, NACE Paper 408, NACE, Houston, Tx, 1992.

13. T. Rogne, J. M. Drugli and R. Johnsen, Testing for Initiation of the Crevice Corrosion of Welded Stainless Steels in Natural Sea water, *Mat. Perform.*, September 1987.

14. T. Rogne, J. M. Drugli and R. Johnsen, Corrosion Testing of Welded Stainless Steel in Sea Water, *Corrosion '86*, Paper 230, NACE, Houston, Tx, 1986.

15. A. Mollica and A. Trevis, The Influence of Microbiological Film on Stainless Steels in Natural Sea Water, *4th Int. Congr. on Marine Corrosion and Fouling*, Juan-les-Pins, 1976.

16. V. Scotto, R.Di Cinitio and G. Marcenaro, The Influence of Marine Aerobic Microbial Film on Stainless Steel Corrosion Behaviour, *Corros. Sci.*, 1985, **25**, 3.

17. J. P. Adudonard, A. Desestret, L. Lemoine and Y. Morizur, Special Stainless Steels for use in Sea water, UK CORROSION, Wembley 1984. Published by the Institute of Corrosion, Leighton Buzzard, UK.

18. S. C. Dexter and G. Y. Gao, Effect of Sea water Biofilms on Corrosion Potential and Oxygen Reduction of Stainless Steels, *Corrosion '87*, Paper 377, NACE, Houston, Tx, 1987.

19. P. Gallager, R. E. Malpas and E. B. Shone, Corrosion of Stainless Steels in Natural, Transported and Artifical Sea waters, *Brit. Corros. J.*, 1988, **23**, (4).

20. A. Mollica, A. Trevis, E. Traverso, G. Ventura, G. De Carolis and R. Dellepiane, Cathodic Performance of Stainless Steels in Natural Sea water as a Function of Microorganism Settlement and Temperature, *Corrosion*, 1989, **45**, (1).

21. V. Scotto, G. Alabiso and G. Marcenaro, *Bioelectrochemistry and Bioenergetics*, BEBO 0945, 1986.

22. U. Steinsmo, to be published. SINTEF Corrosion and Surface Technology, Trondheim, Norway, 1995.

23. D. Thierry *et al.*, Effect of Marine Biofilms on Stainless Steels — Results from a European Exposure Programme, 1995 Int. Conf., MIC, New Orleans, May, 1995.

24. F. Mansfeld and B. Little, A Technical Review of Electrochemical Techniques Applied to Microbiologically Influenced Corrosion, *Corros. Sci.*, 1991, **32**, (3), 247.

25. J. M. Drugli and T. Rogne, Cathodic Properties of Stainless Steel in Chlorinated Sea Water, SINTEF report STF 24 F95378, 1995-11-24.

26. NORSOK STANDARD — Materials Selection, M-DP-001, Rev. 1. December, 1994.

27. R. Johnsen, Corrosion Failures in Sea Water Piping Systems Offshore, Marine and Microbial Corrosion, 1991, Stockholm, Sweden.

28. B. Wallen and S. Henrikson, Effect of Chlorination on Stainless Steels in Sea water, *Corrosion '88*, Paper 403, NACE, Houston, Tx, 1988.

29. R.M. Kane and P.A. Klein, Crevice Corrosion Propagation Studies for Alloy N06625, Remote Crevice Assembly Testing in Flowing Natural and Chlorinated Sea water, *Corrosion '90*, Paper 158, NACE, Houston, Tx, 1990.

30. J. M. Drugli, Method and Arrangement to Hinder Local Corrosion and Galvanic Corrosion in Connection with Stainless Steels and other Passive Materials, Patent claim 910993, revised 1994-06-10.

31. P. O. Gartland and J. M. Drugli, Methods for Evaluation and Prevention of Local and Galvanic Corrosion in Chlorinated Sea Water Pipelines, *Corrosion '92*, Paper 408, NACE, Houston, Tx, 1992.

32. R. Johnsen, P. O. Gartland, S. Valen and J. M. Drugli, The Resistor Controlled Cathodic Protection Method for Prevention of Local Corrosion on Stainless Steels in Saline Water Piping

Systems, paper presented at the *7th Middle East Corrosion Conf.*, Bahrain, February, 1996.
33. R. Johnsen, P. O. Gartland, S. Valen and J. M. Drugli, Internal Cathodic Protection of Sea Water Piping Systems by use of the RCP Method, *Corrosion '96*, Paper 559, NACE, Houston, Tx, 1996.
34. P. O. Gartland, R. Johnsen, S. Valen, T. Rogne and J. M. Drugli, How to Prevent Galvanic Corrosion in Sea Water Piping Systems, *Corrosion '96*, Paper 496, NACE, Houston, Tx, 1996.
35. I. H. Hollen, Sea Water Systems-Experiences from Draugen, European Workshop on Sea Water Corrosion of Stainless Steels — Mechanisms and Experiences, Published by SINTEF Corrosion and Surface Technology, N-7034 Trondheim, Norway.

6

Corrosion Behaviour of Stainless Steel in Thermally Altered Sea Water

D. FÉRON

CEA/CEREM-SCECF - BP6 - 92265, Fontenay-aux-Roses, France

1. Introduction

During the last decade, a great deal of research has been conducted to understand the behaviour of stainless steels in natural sea water. The European project "Marine biofilms on stainless steels", partially founded by the European Community, has contributed to the increase of knowledge in this field. One aspect of this project focused on stainless steel behaviour in thermally altered sea water, since sea water is often used in heat exchanger systems. The results of this particular research deal with sea water at temperatures between ambient and 40°C, the influence of sea water temperature on both biofilm growth or properties, and on the corrosion resistance of stainless steels.

2. Experimental

The main experimental procedures (both microbiological and biochemical) have been previously reported [1, 2]. Only specific configurations related to the tests performed with thermally altered sea water are reported here.

2.1. Materials

Ten different stainless steel (SS) grades of European production, mainly high grade materials, were used. They included austenitic and duplex steels and alloys containing 2–6% molybdenum (Table 1). Specimens were mainly in the form of tubes (Ø 23/25 mm, length 100 mm) and plates (60 × 60 × 3 mm).

Crevice corrosion specimens were made from stainless steel plates with crevice formers made of polyoxymethylenechloride (POM) rings and fixed with 254 SMO bolts and nuts, the applied torque being 3 Nm. The metal part of the crevice former was electrically isolated by a PTFE tape. Those parts of the plate where the crevice former was applied, were machined in order to present the same roughness whatever the steel grade.

Before exposure, all the specimens were pickled at ambient temperature for 20 min in a solution containing 20% HNO_3 and 2% HF.

Table 1. Chemical composition of tested stainless steels (wt %)

Alloy	C	Ni	Cr	Mo	N
URANUS B26	0.009	24.7	20	6.3	0.19
URANUS SB8	0.01	25	25	4.7	0.21
URANUS 47N	0.01	6.6	24.7	2.9	0.18
URANUS 52N+	0.01	6.3	25	3.6	0.25
654 SMO	0.01	21.8	24.5	7.3	0.48
254 SMO	0.01	17.8	19.9	6	0.2
AISI 316	0.017	12.6	17.2	2.6	0.05
SAF 2507	0.14	6.9	24.9	3.8	0.28
SAF 2205	0.15	5.5	22	3.2	0.17
SAN 28	0.019	30.3	26.7	3.4	0.07

2.2. Exposure Conditions

All tests were performed at the CEA sea water facility, called the SIRIUS facility, located at La Hague (Normandy, France). The sea water was pumped at high tide from the rocky coast into reservoirs of 120 m^3 maximum volume. There is no access of light to the sea water between the pumping system in the sea and the exposed coupons. In the SIRIUS facility, the sea water is heated by a series of titanium heat exchangers and is completely renewed: it is a " single pass " loop, and there is no recirculation of the sea water. Tubes and titanium autoclaves are put between two heat exchangers, at the desired temperature (Fig. 1).

The sea water flows at 1.5 $m^3 h^{-1}$ inside the stainless steel tubes (1 ms^{-1}) and inside titanium autoclaves (0.01 ms^{-1}) as shown in Fig. 1. When titanium autoclaves are used, sea water is heated to 40°C in 15 min, but when there is no autoclaves, the time to heat the sea water to 40°C is less than 2 min.

2.3. Measurements

The free corrosion potentials of tubes and plates were measured vs saturated calomel electrode (SCE) every hour by means of a Solartron Schlumberger data acquisition system (3351 D ORION).

Sea water temperatures, pH, redox potentials (platinum potential vs SCE) were also recorded every hour.

Chemical analyses of the sea water were conducted during the tests and included analyses of the main cations and anions, dissolved oxygen, total conductivity, organic matter content and chlorophyll a. Mean values obtained during the reported tests are shown on Table 2. No major variations were observed, except the ambient temperatures of the sea water which varied from 7°C in winter up to 17°C in summer.

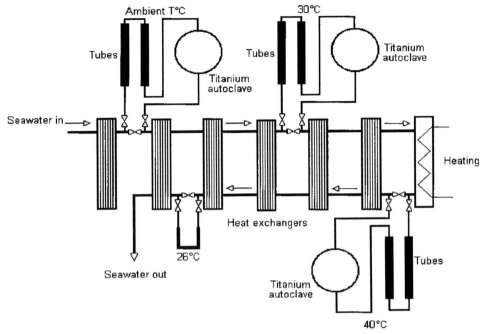

Fig. 1 *Schematic diagram of the SIRIUS facility.*

Table 2. *Sea water composition during the MAST II tests (mean values with s.d.)*

C'vity mS cm^{-1}	pH	Cl$^-$ g kg^{-1}	SO$_4^{2-}$ g kg^{-1}	Na$^+$ g kg^{-1}	K$^+$ g kg^{-1}	Mg^{2+} g kg^{-1}	Ca^{2+} g kg^{-1}	O$_2$ mg kg^{-1}	Chloro a µg kg^{-1}	Organic mg kg^{-1}
40.0 (2.5)	8.0 (0.1)	18.0 (1.2)	2.7 (0.)	10.8 (0.2)	0.5 (0.1)	1.7 (0.2)	0.7 (0.1)	7.9 (0.7)	0.3 (0.2)	7.0 (3.0)

3. Results and Discussion
3.1. Free Corrosion Potential Evolution

Experiments were conducted with twenty tubes made of superaustenitic steel (654 SMO) at each temperature. Free corrosion potential evolutions are illustrated in Fig. 2 for these tubes and may be summarised as follows :

- at 20°C, previous results obtained at ambient temperature are confirmed [1, 2], i.e. a rapid increase of the corrosion potentials of the SS tubes up to +300 mV(SCE) in about 10 days
- at 30°C, a similar shift of the SS potentials up to +300 mV(SCE) was also observed, perhaps a little slower, in about 12 days
- at 40°C, the free corrosion potentials of all the tested SS tubes remained constant at about –100 mV(SCE) which corresponded to the starting value of the potentials at 20 and 30°C.

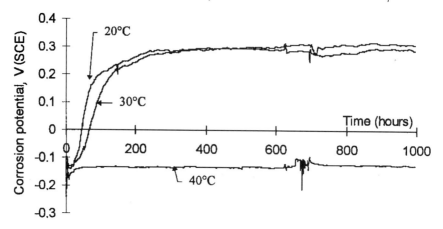

Fig. 2 *Free corrosion potential evolution of stainless steel tubes (flowing natural sea water, 1 ms⁻¹).*

There is no difference between these results and those obtained with other superaustenitic tubes (URANUS 52N) which were exposed at 20 and 40°C.

The only difference between the E_{corr} evolution on tubes and plates was in the rate of the potential increase — which was always faster on plates than on tubes and probably correlated with the sea water flow velocity being slower with the plates.

These results are in agreement with other published results [3–5] in which potential rises have been reported to occur only at temperatures below 30–40°C, the exact temperature limits depending to some extent on the geographical location or on the temperature differences between the ambient temperature and the heated values.

As shown in Fig. 3, where E_{corr} of plates are reported after 30 days exposure, differences between the free corrosion potentials of titanium occur between 20 and 40°C. At 20°C, titanium E_{corr} values are ≈250 mV, at 30°C, ≈150 mV and at 40°C, ≈10 mV(SCE). The same phenomena probably take place on stainless steel and on titanium specimens at the same temperatures and so it is related to the biofilm characteristics. On platinum, high potential values are obtained whatever the temperature, while the starting values of the redox potentials are nearly the same at 20, 30 or 40°C — i.e. between 100 and 200 mV(SCE).

In another series of tests, five superaustenitic tubes were exposed to sea water heated at 30°C and five other tubes were located in the section where the same sea water was cooled to 25°C after being heated up to 40°C. The E_{corr} evolution of these tubes is presented on Fig. 4; at both temperatures, with sea water heated to 30°C or with sea water cooled from 40 to 25°C, there was an increase of E_{corr}, although faster at 30°C. A temporary heating to 40°C does not prevent the E_{corr} increase of the sea water.

Other tests have shown that if a biofilm is first formed on SS tubes at 30°C and then these are exposed at 40°C, the E_{corr} values at 40°C are stable at ≈250 mV(SCE) (starting values at 40°C) during 10 days and then decrease to ≈–50 mV(SCE) in a further 5 days. So even if a biofilm has previously formed at lower temperatures, the free corrosion potential at 40°C will fall to a stable low value, below 0 mV(SCE).

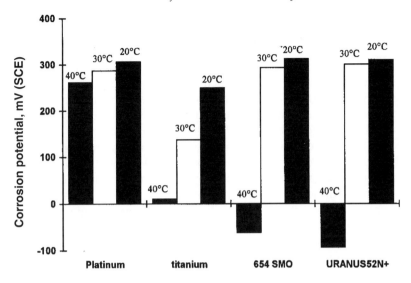

Fig. 3 *Comparison of the free corrosion potentials of platinum, titanium and stainless steels after 30 days exposure.*

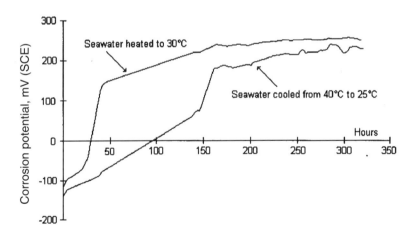

Fig. 4 *Free corrosion potential evolution of stainless steel (654 SMO) with heated and cooled sea water.*

3.2. Biofilm Investigation

Sampling of biofilms was performed on tubes exposed at the three temperatures in order to characterise the biofilms formed on these tubes. First investigations were on biofilm mass and on exopolymeric substances, called EPS*. A correlation was also

---

*EPS analyses were performed by Dr. V. Scotto at ICMM, Genova (Italy).

found between the E_{corr} and the biofilm EPS developed on SS surfaces at ambient temperatures in all the European stations involved in the Biofilm European program [2].

Table 3 summarises the results obtained on tubes exposed at 20, 30 and 40°C : biofilms were present on all the SS tubes, including those at 40°C, and large amounts of EPS were found also at 40°C. The fact that there is no ennoblement of E_{corr} at 40°C, in spite of a massive presence of EPS, shows that the exopolymeric substances are not responsible for the electrochemical effects of the biofilms.

More information on the biofilm settlement was obtained by direct epifluorescence on exposed SS surfaces. During experiments performed in the SIRIUS facility, direct observations of micro-organisms on surfaces by epifluoresence showed that bacteria are numerous on SS surfaces, and therefore difficult to count. In a first attempt, the biofilm settlement on SS surfaces was evaluated by the percentage of the surface covered by micro-organisms (dead or alive), and eventually by substances which react also with the fluorescent substance (DAPI). These observations were conducted at different exposure times and at different E_{corr} values. The results obtained on coupons exposed at 20 and at 40°C are summarised in Table 4. They show clearly that the percentage of the SS surfaces covered by micro-organisms is the same or even more at 40°C than at 20°C [7] while the SS free corrosion potential stays constant at 40°C but increases at 20°C. This clearly shows that the presence of a biofilm on SS surfaces is not in itself responsible for the potential ennoblement. This could perhaps be seen in relation to some published results, where no ennoblement of SS in natural sea water was reported [6].

Investigations were then made of the number of bacteria that were present in the biofilm, to determine if there was a major difference between the bacterial colonisations at 20°C and at 40°C which could explain the electrochemical behaviour. Aerobic, facultative anaerobic and strict anaerobic bacteria were isolated and counted[†]. As shown in Table 5, there was no major difference between the number of bacterial populations.

The more important differences between bacterial colonisations at 20°C and at 40°C were obtained by numerical taxonomy[†]. Specific "phenons" (bacterial genus or species) are found at 20°C and not at 40°C : all isolated bacteria are gram-negative, but bacteria found on stainless steel surfaces at 40°C do not grow on sugars, including glucose, or on "polyols". These results indicate that the metabolic or the enzymatic

Table 3. Biofilm sampling on SS tubes exposed to sea water thermally altered

Temperature		20°C		30°C		40°C	
Duration (day)	5	27	5	27	13	27
Potential (mV(SCE))	250	315	200	300	−90	−120
Wet biofilm ($mgcm^{-2}$)	0.22	14	0.32	11	0.74	4.4
Protein (μgcm^{-2})	1.6	46	3.4	52	4.4	9.9
EPS (μgcm^{-2})	98	3431	—	2871	—	1862

[†] These analyses were performed by E. Jacq and S. Corre at MICROMER, Brest (France).

Table 4. Biofilm settlement on stainless steel surfaces (epifluorescence observations of the percentages of the SS surface covered with bacteria)

Exposure	20°C		40°C	
time	Potential	Covered surface	Potential	Covered surface
(h)	mV (SCE)	(%)	mV(SCE)	(%)
24	110	5	−120	4
48	0	11	−135	20
72	200	10	−115	40
96	250	20	—	—
144	—	—	−140	40
240	290	25	—	—
312	—	—	−140	41

Table 5. Number of bacteria on stainless steel surfaces (after 200 h sea water exposure)

Exposure temperature	Total number of bacteria [1]	Marine heterotroph bacteria [2]		
		Aerobic	Facultative	Anaerobic
20	$4.4 \ 10^7$	$6.0 \ 10^4$	$2.3 \ 10^1$	<0.75
40	$3.3 \ 10^7$	$1.9 \ 10^5$	0.9	2.3

[1] By epifluorescence (number of bacteria cm^{-2}).

[2] By MPN method.

activity of bacteria which are in the biofilm form the parameter which controls the potential ennoblement [9].

Some bacteria found at 20°C grew on glucose, so the enzymatic activity was tested with the enzyme glucose oxidase, which catalyses the following reaction:

$$\text{glucose} + \text{oxygen} \rightarrow \text{gluconic acid} + \text{hydrogen peroxide}$$

When this glucose oxidase was added to sterile and aerated sea water containing some glucose, potential ennoblement was obtained on stainless steel coupons as illustrated in Fig. 5. It is significant that the final E_{corr} values are in accordance with those obtained during exposure to natural sea waters. This ennoblement with glucose oxidase is due to the formation of organic acid and of hydrogen peroxide which accelerates the rate of the cathodic reaction [8,9]. Such a result can be correlated with previously reported experiments, which demonstrated the presence of hydrogen peroxide in biofilms developed on platinum exposed to natural sea water [5].

3.3. Crevice Corrosion

Crevice corrosion test with artificial crevice formers were carried out on ten SS grades exposed in natural sea water heated in a single pass loop using a series of heat

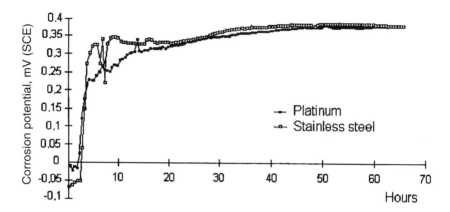

Fig. 5 *Evolution of the free corrosion potential of stainless steel in sterile sea water with addition of glucose oxidase at 30°C.*

exchangers in the SIRIUS facility. Five specimens of each grade were tested at each temperature (20°C, 30°C, 40°C).

The corrosion potentials of non-corroded specimens were about 300 mV(SCE) at 20 and 30°C (Fig. 6). The values were between –50 and –100 mV(SCE) at 40°C. At 20 and 30°C, corroded specimens had lower corrosion potentials and fluctuated, as illustrated by the 316 L potential behaviour in Fig. 6. At 40°C, all the potentials were in the same range, but more fluctuating for specimens on which depassivation was observed after the test.

All the stainless steel grades, excluding the 316 L grade, showed a very good behaviour at the three test temperatures: only few examples of depassivation —

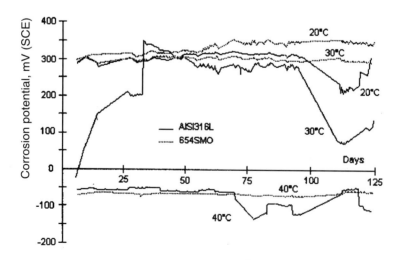

Fig. 6 *Potential evolution of crevice coupons exposed to flowing sea water.*

revealed as changes in colour — were observed under the gaskets but without weight losses or depth of penetration.

Three of five 316 L specimens showed through-wall corrosion at 20°C and 30°C. Only a "depassivation" (colour change, no weight losses, no depth) occured on 316L specimens at 40°C. This temperature effect can be directly correlated to the biofilm and the potential evolution; thus, biofilm formation on stainless steel surfaces leads to a high corrosion potential at 20°C and 30°C, and this promotes crevice corrosion. Biofilm formation also at 40°C occurs but without increase of the corrosion potential, and a better behaviour of 316 L is observed at 40°C even when corrosion is thermally activated.

4. Conclusions

An increase of the free corrosion potential of stainless steels in natural sea water occurs at 20°C and 30°C, the potential rising from about –100 mV to some +300 mV(SCE). At 40°C, the free corrosion potentials of stainless steels are constant at about –100 mV(SCE). Biofilm formation and bacterial settlement on stainless steel surfaces occur at the three test temperatures.

The biofilms that developed at the three test temperatures do not present differences in terms of global parameters (masses, bacterial numbers, polysaccharide contents, etc). The only differences were in the metabolic or the enzymatic activities of the bacteria. The hydrogen peroxide generated as the metabolic reaction of bacteria (enzymatic reaction with glucose oxidase) explains the electrochemical effects at 20 and 30°C.

Crevice corrosion results are in accordance with the potential evolution obtained at 20 and 30°C: crevice corrosion occurs on all 316 L specimens, while depassivation is only observed at 40°C. Good behaviour is observed for all the other tested stainless steels: 654 SMO, 254 SMO, SAF 2507 and 2205, SAN 28, URANUS B26, SB8, 47N and URANUS 52N+.

All these results agree with a decrease of the corrosivity of natural sea water as far as stainless steels are concerned when temperature increases from ambient to 40°C. This decrease of sea water corrosivity could be due to changes of the metabolic properties of the biofilms, such as the enzymatic production of acids and hydrogen peroxide by bacteria.

References

1. J. P. Audouard *et al.*, in *Proc. 3rd EFC Workshop on Microbial Corrosion*, Estoril, Portugal, 1994. European Federation of Corrosion Publications No. 15, The Institute of Materials, London, UK (1995).
2. J. P. Audouard *et al.*, 1995, in *Proc. Int. Congr. on Microbial Induced Corrosion*, New Orleans, USA, NACE, Houston, Tx.
3. A. Mollica and A. Trevis, in *4th Int. Congr. on Marine Corrosion and Fouling*, Juan-les-Pins, France, 1976.

4. E. Bardal, J. M. Drugli and P. O. Gartland, *Corros. Sci.*, 1993, **35**, (1–4) 257–267.
5. P. Chandrasekaran and S. C. Dexter, *Corrosion '93*, paper No. 493, NACE, Houston, Tx, USA, 1993.
6. B. J. Little and F. Mansfeld, *Corrosion '90*, paper No. 150, NACE, Houston, Tx, USA, 1990.
7. I. Dupont, G. Novel and D. Féron, *4ème Congr. national de la Société Française de Microbiologie*, Tours, France, 1995, Paper VE-14.
8. H. Amaya and H. Miyuki, *Corros. Engng*, 1995, **44**, 123–133.
9. I. Dupont, G. Novel and D. Féron, to be presented at EUROCORR '96, Nice, France, 1996.

7

Statistical Approach of Pitting Initiation of Stainless Steel in Flowing Sea Water

C. COMPÈRE, P. JAFFRÉ and L. ABIVEN

IFREMER, Centre de Brest, BP 70, 29280 Plouzané, France

ABSTRACT

Localised corrosion is governed by random parameters and analysed according to the stochastic theory. The influence of ageing in sea water and of the open-circuit potential values of stainless steels on the breakdown potential distribution and on the elementary pitting probability ϖ per unit area, determined under potentiokinetic conditions at $v = 1.67$ mV s^{-1} is studied.

The role of the scanning rate on the breakdown potential distribution is also examined. A competition phenomenon, dependent on the anodic polarisation scanning rate, seems to appear between pit generation and secondary passivation for long periods of immersion and high open-circuit potential values.

1. Introduction

In sea water the most widely used 316L grade stainless steels suffer from localised corrosion and are rapidly subject to crevice corrosion. Localised corrosion is governed by random parameters ,and for this reason may be considered under a probabilistic aspect. The stochastic approach put forward a few years ago by Shibata and Takeyma [1, 2], has been used in the present study. In this, the distribution of the breakdown potentials was analysed as a function of ageing time in sea water and of the open-cicuit potential values. The importance of these two parameters on the elementary pitting probability per unit area, ϖ, determined under potentiokinetic conditions, was also evaluated from a statistical point of view.

Numerous studies have been performed to obtain pitting potentials under different experimental conditions [3, 4], but in this paper a particular emphasis was given to the anodic scanning rate of potential.

2. Experimental Procedure

The composition of the commercial AISI 316L stainless steel samples given in Table 1 was measured using glow discharge lamp (GDL) excitation. The samples were disks of 16 mm dia. embedded under vacuum in an epoxy resin. This technique is

Table 1. Composition of the AISI 316L stainless steel

C	Mn	Si	P	S	Cr	Mo	Cu	Al	Ni	Ti
0.023	1.19	0.45	0.027	0.017	16.40	1.98	0.46	0.01	11.01	0

particularly suitable for reducing the risks of crevice corrosion at the interface between the resin and the specimen. The experimental assembly consists of three water lines in parallel with a by-pass line. Each water line has two test cells in series, each containing eight specimens, as shown in Fig. 1. One Ag/AgCl reference electrode in sea water was installed into each test cell. The eight specimens and a platinum coated titanium wire used as counter electrode were installed parallel to the flow direction. Sea water is pumped from the Brest Bay (France), roughly filtered and stored in a water tower before being used for feeding the experimental set-up in the laboratory. The natural sea water flow rate was 1 m s^{-1}. Prior to the test, the samples were ground with SiC emery paper (final stage 1200 grit) then pickled at ambient temperature for 20 min in a 20 vol.% solution of HNO_3 (65 wt%) and 2 vol.% solution of HF (50 wt%), rinsed in deionised water, immersed for 20 min in a 20 vol.% solution of HNO_3 (65 wt%), rinsed in deionised water and finally air-aged at ambient temperature for 24 h.

The pitting resistance of stainless steels samples was studied by conducting, in potentiokinetic mode, anodic polarisation curves as a function of immersion time in natural sea water. Polarisation was done from the open-circuit potential up to a critical value of anodic current of 0.1 mA cm^{-2} at a scan rate of 1.67 mV s^{-1}. Some complementary tests were performed at 0.833 and 0.0833mV s^{-1}. The breakdown potential was estimated from an anodic current of 0.08 mA cm^{-2}. This value was

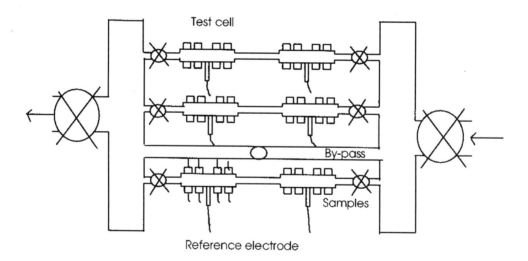

Fig. 1 Experimental assembly.

carefully chosen in order to characterise the onset of stable pitting. The comparison of the beakdown potential values determined from a constant anodic current density is possible since the slopes of the polarisation curves, when the current increases as a function of potential, were similar for all measurements. After each polarisation, the samples were observed with a microscope with a particular attention given to the resin/stainless steel interface. Samples presenting pitting corrosion only were considered for the statistical analysis while crevice corroded samples were rejected. The electrochemical device was a multi-channel potentiostat (POLUCOR developed by ORCA, Plouzané, France). The different tests are summarised in Table 2. The number of experiments generally exceeded 55 at each exposure time except for test periods longer than 60 d.

3. Results
3.1. Theoretical Analysis

A pitting probability per unit area ϖ and a pit generation rate g were first introduced by Baroux *et al.* [5]. The surface area of the sample S is divided into small elementary parts δS in the case of an homogeneous surface, the survival probability P satisfies the following relations:

$$P = (1 - \varpi \delta S)^{\frac{S}{\delta S}} \tag{1}$$

$$\lim_{\delta S \to 0} P = \exp(-\varpi S) \tag{2}$$

Therefore:

$$\varpi = \frac{-1}{S} LnP = \frac{-1}{S} Ln \frac{i}{N} \tag{3}$$

The survival probability P may be estimated using the total number of samples N and the number of non-pitted samples i. The results are expressed in terms of an elementary pitting probability per unit area ϖ.

3.2. Distribution of Pitting Potential as a Function of Ageing and Open-circuit Value

It was shown in a previous paper [7] that the free corrosion potential of passive stainless steels rapidly increases as a function of immersion time in natural sea water, which is the expected behaviour in such experimental conditions [6]. Since the biofilm growth seems to have an important role in this ennoblement, the effect of composition,

Table 2. Total number of tests as a function of immersion time

Immersion time (days)	2	5	9	14	21	> 60
Total number of samples	56	60	56	55	50	19

thickness and ageing of passive films in sea water must not be neglected [8]. The free corrosion potential increases near the breakdown potential values and enhances the probability of pitting corrosion. However, the pitting potential also changes as a function of ageing time. The histogram in Fig. 2 illustrates the influence of the immersion time on the distribution of pitted samples over the breakdown potential ranges. A peak in distribution is observed after 2, 5, and 9 d of immersion around 515 mV (SCE). After 14 and 21 d of immersion, no peak is observed anymore since the pitted samples are evenly distributed over the whole potential range. A peak in distribution is again observed for long periods of immersion, above 60 d, in natural sea water at very high potential values around 1100 mV (SCE) . In view of the interdependence between free corrosion potential and immersion time, it seems interesting to analyse the distribution of pitted samples vs five open-circuit potential ranges, as indicated in Table 3. The histogram showing the percentage of pitted samples and the repartition of the breakdown potential for these five open-circuit potential ranges is given in Fig. 3. A peak in distribution is again observed for the first four ranges around 500 mV(SCE) but for an open-circuit potential value higher than 65 mV (SCE), the number of pitted samples is uniformly distributed along a larger range of pitting potentials. Thus, a long period of immersion in sea water and an open-circuit potential higher than 65 mV (SCE) seem to define a completely different pitting resistance behaviour.

The random variation of the breakdown potential about its average value follows a normal distribution as shown in Figs 4 and 5 in the case of analyses as a function of

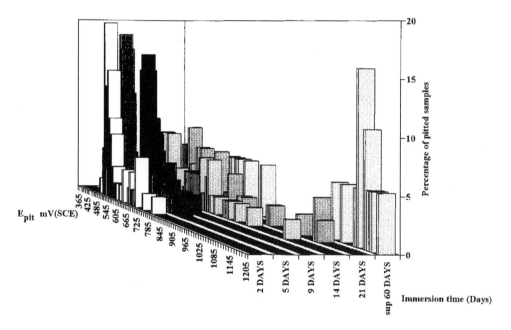

Fig. 2 Influence of the immersion time on the distribution of pitted samples over the pitting potential ranges.

time as well as of open-circuit potential. As the immersion time increases the curves evenly flatten, become wider and shift towards higher potentials indicating a dispersion and a tendency to the ennoblement of the breakdown potentials. All curves are nearly superimposed for open-circuit potential ranges lower than 65 mV(SCE), while for the range higher than 65 mV(SCE), the behaviour is completely different having a maximum of pitting probability around a higher potential value. These results again reveal that a critical open-circuit potential value, around 65 mV (SCE), plays a significant role in enhancing the pitting corrosion resistance.

3.3. Pit Generation as a Stochastic Process

Figures 6 and 7 give the elementary pitting probability as a function of potential for

Table 3. Distribution of the total number of tests as a function of open-circuit potential ranges

Open-circuit potential range mV (SCE)	< -79	$-79 < E_{oc} < -34$	$-34 < E_{oc} < 0$	$0 < E_{oc} < 65$	> 65
Range name	S1	S2	S3	S4	S5
Total number of samples	63	58	58	61	88

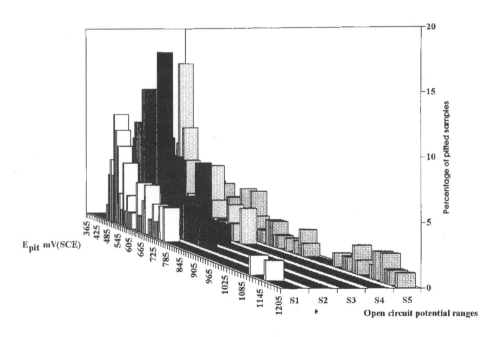

Fig. 3 Influence of the open circuit potential ranges on the distribution of pitted samples over the pitting potential ranges.

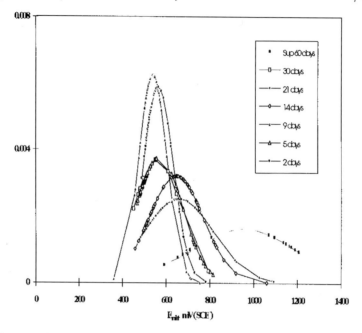

Fig. 4 *Gaussian distribution function of breakdown potentials as a function of time.*

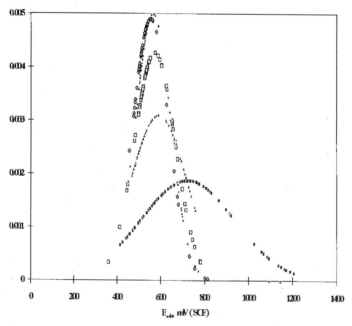

Fig. 5 *Gaussian distribution function of breakdown potentials as a function of open-circuit potential ranges.*

the different immersion times and the different open-circuit potential ranges; they indicate the presence of several stochastic processes. The elementary pitting probability ϖ cannot be approximated by an exponential function as in the works of Baroux [5]. These curves confirm the marked effects of immersion time and open-circuit potential values on the distribution of the breakdown potentials. If we consider an elementary pitting probability per unit area of 0.5, it is clear that the breakdown potential strongly increases with immersion time. Figure 7 indicates that the elementary pitting probability per unit area do not gradually evolve with the open-circuit potential values but sharply changes with a particular potential value.

3.4. Effect of Anodic Potential Scanning Rate

Significant differences can be noted in the breakdown potential evaluation depending on the experimental conditions and particular attention must be paid to this point. The influence of anodic potential scanning rate has been previously studied by Baroux [5]. A critical value of scanning rate has to be taken into account (between 0.75 and 1.67 mV s^{-1} for 304L stainless steel in NaCl solution) under which breakdown potentials have to be considered with care. The results of potentiokinetic tests are

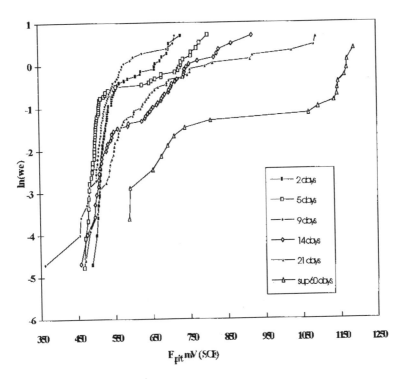

Fig. 6 *Elementary pitting probability ϖ as a function of the potential for the different immersion tests.*

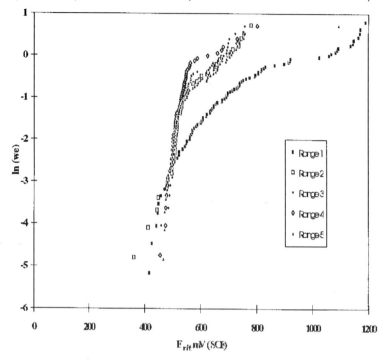

Fig. 7 *Elementary pitting probability* ϖ *as a function of the potential for the different open circuit potential ranges.*

undoubtedly dependent on the scanning rate v as indicated in Figs 8(a–c) giving the anodic polarisation curves at v =1.67, 0.83 and 0.083 mV s^{-1} respectively. These curves were obtained after 48 days of immersion. After a long immersion time a competition phenomenon appears between pit generation and secondary passivation depending on the scanning rate. These results explain the extremely high values of breakdown potentials, around 1100 mV (SCE), for long immersion time and for high value of open-circuit potential. The higher the scanning rate is, the more evident the secondary passivation becomes. These results are not completely explained so far and complementary experiments and surface analyses are still required.

4. Conclusions

The following conclusions can be drawn from the results discussed above:

1. The breakdown potentials expressed as a function of time or open-circuit potential values follow the Gaussian law of distribution. The breakdown potential distribution may be described by the peak height and the standard deviation.

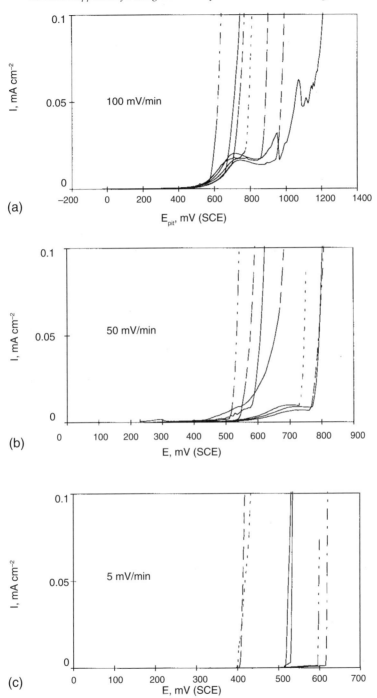

Fig. 8 *Anodic polarisation curves obtained on 316L stainless steel after 48 days of immersion time in sea water: (a) scan rate* $v = 1.67\ mV\ s^{-1}$; *(b)* $v = 0.833\ mV\ s^{-1}$; *(c):* $v = 0.083\ mV\ s^{-1}$.

2. Time of ageing in sea water and open-circuit potential values have marked effects on the elementary pitting probability per unit area ϖ determined under potentiokinetic conditions at $v = 1.67$ mV s^{-1}.

3. A competition phenomenon appears between pit generation and secondary passivation for long periods of immersion and high open-circuit potentials depending on the anodic polarisation scanning rate.

References

1. T. Shibata and T.Takeyma, *Corrosion*, 1977, **33**, 7, 243.
2. T. Shibata and Y. Zhu, *Corros. Sci.*, 1994, **36**, 1, 153.
3. G. Salvago and G. Fumigalli, *Corros. Sci.*, 1994, **36**, 4, 733.
4. G. Salvago and G. Fumigalli, *Corros. Sci.*, 1995, **37**, 8, 1303.
5. B. Baroux, *Corros. Sci.*, 1988, **28**,10, 969.
6. C. Compère, P. Jaffré and D. Festy, *Corrosion*, 1996, **52**, 7, 496.
7. J. P. Audouard, C. Compère, N. J. E. Dowling, D. Féron, D. Festy, A. Mollica, T. Rogne, V. Scotto, U. Steinsmo, K. Taxen and D. Thierry, *Proc. 3rd Int. EFC Workshop* (eds. A. K. Tiller and C. A. C. Sequeira), 1995.
8. C. Compère, B. Rondot, M. G. Walls and M. da Cunha Belo, submitted to EUROCORR '96, Nice, 24–26 September 1996.

8

Stainless Steel Corrosion in Antarctic Sea Waters: A Contribution from the Italian PNRA Project

G. ALABISO*, V. SCOTTO, A. MOLLICA, U. MONTINI, G. MARCENARO and R. DELLEPIANE

Istituto per la Corrosione Marina Dei Metalli (ICMM-CNR), Via de Marini 6, 16149 Genova, Italy
*Istituto SperimentaleTalassografico "A. Cerruti" (ISTTA-CNR), Via Roma 3, 74100 Taranto, Italy

ABSTRACT

The main scientific goal of this study was to establish if biofilms, grown in Antarctic conditions, are able to modify the oxygen reduction curve on stainless steel surfaces as in the European Seas.

The experimental activity was carried out at the Italian Base located in Terra Nova Bay (Ross Sea, Victoria Land) and consisted of short and long term tests.

The short term tests, which were run in the laboratory with plain continuously renewed sea water during the summertime stay, aimed to collect electrochemical data (such as polarisation curves, open circuit potentials, galvanic currents, etc) and biofilm samples for subsequent analytical characterisation.

The long term tests aimed to monitor in the sea the corrosion behaviour of a wide range of stainless steels for marine use. In these tests a loading structure with about a hundred specimens was immersed at a depth of 50 m in Adelie Cove and recovered after 1 and 3 y.

Biofilm adhesion on surfaces modifies the oxygen reduction modality also in Antarctica but, in contrast with the European Seas, the open circuit potentials of passive stainless steels never exceed +50 – +100 mV (SCE). As a consequence, the risk of onset of localised corrosion is lower; however cathodic curves and weight losses suggest that the propagation rate of localised attack is increased by biofilm growth as in European Seas.

1. Introduction

The Italian Research activity in Antarctica started in the middle of the 1980s with the settlement of a Base in Terra Nova Bay, (Ross Sea, Victoria Land) Lat.: 74°41′42"S; Long.: 164°07′23"E (Fig. 1). The research activity was supported by the Italian National Antarctic Research Programme (PNRA) and was devoted at first to the hydrogeological characterisation of the region.

Later on, a part of the activity focused on the technological problems that low temperatures could cause and the study of the corrosion behaviour of alloys of wide technological interest in the Antarctic conditions was one of the aspects taken into account.

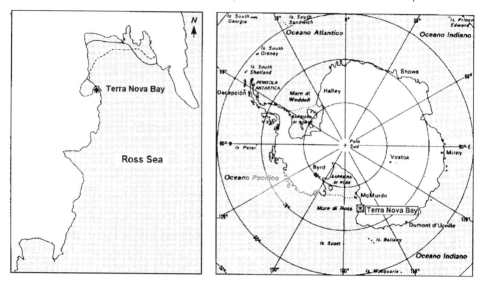

Fig. 1 The Italian summer station at Terra Nova Bay, Ross Sea, Antarctica.

It is well known from the literature [1–11] and from the activity carried out in the MAST Project BIOFILM [12– 14], that the adhesion of a marine biofilm on surfaces depolarises the oxygen reduction modalities with three main consequences:

–   increase in the free corrosion potentials of stainless steels in the passive state up to +300 – +400 mV (SCE) so increasing the probability of the onset of localised corrosion.
–   increase the localised corrosion propagation rates.
–   increase the galvanic currents of less noble materials when coupled with stainless steels covered by biofilms.

The main scientific goal of this study was to establish if marine biofilms grown in Antarctic conditions could have the same effects on stainless steels and on active passive alloys as has been observed in the European Seas.

2. Materials and Methods

The activity carried out in Antarctica during the 1989/90, 1990/91, 1993/94 and 1994/95 campaigns consisted of short and long term tests.

2.1. Short Tests

The short term tests were carried out at the Italian Base in Terra Nova during the summertime stay and lasted no more than 2 months. The tests were performed in tanks of continuously renewed plain sea water maintained at a temperature close to the sea (that is 2°C instead of –1/–2°C).

The activity aimed to:

– measure the evolution in time of the open circuit potentials on passive stainless steels;
– collect biofilm samples for their subsequent analytical characterisation; and
– carry out potentiostatic polarisation tests to follow in time the modifications produced in the oxygen reduction modalities by biofilm settlement on stainless steel surfaces.

The detachment of biofilm from surfaces was carried out in a buffer containing EDTA by means of glass balls. After centrifugation at 9000 rpm the obtained pellet was freeze dried and the supernatant liquid frozen at –20°C.

The analytical characterisation of biofilms was subsequently conducted in Italy.

2. 2. Long Tests

The long term activity aimed to follow the corrosion behaviour of a wide range of alloys for marine use, with particular attention for stainless steels.

The stainless steel grades tested are reported in Table 1 together with the immersion period. The loading structure was immersed at a depth of 50 m in Adelie Cove and recovered after 1 and 3 years respectively.

The structure was equipped with a data logger for measurements of open circuit potentials and currents delivered in a remote crevice assembly. This last was formed by an anodic area of around 22 cm^2 of each tested alloy coupled, through fixed resistors, to 450 cm^2 of a cathodic area of a high quality stainless steels such as 20Cr–18Ni6Mo. Unfortunately, the data acquisition was compromised by water infiltration and therefore only weight loss measurements were obtained at the end of the field tests.

Table 1. Stainless steel grades tested and immersion period

Type	Immersion date	Recovery date	Immersion period (months)
AISI 304	16/01/90	03/01/91	12
AISI 316	16/01/90	03/01/91	12
AVESTA 18-9	04/02/91	02/12/93	34
AVESTA 17-12-2.5	04/02/91	02/12/93	34
AVESTA 904L	04/02/91	02/12/93	34
AVESTA 254 SMO	04/02/91	02/12/93	34
AVESTA 17-14-4LN	04/02/91	02/12/93	34
AISI 316L	04/02/91	02/12/93	34
AISI 304 L	04/02/91	02/12/93	34

3. Results

The main results obtained during the 1989/90 and 1990/91 campaign have already been published [15, 16].

This paper reports some further information on the long term (3 years) corrosion behaviour of stainless steels and the main results of the analytical characterisation of biofilms.

3.1. Short Term Tests

The mean chemical composition of an Antarctic sea water sample is reported in Table 2 together with salinity, pH, dissolved oxygen content and temperature measured in the field at the exposure depth during the summer time expeditions.

3.1.1. Biofilm Analyses
The Antarctic biofilm samples were analysed in Italy in order to separate out the exopolymeric substances (EPS) amounts and the protein (PRT) and carbohydrate (CHO) contents as indexes of biomasses adhering on surfaces during the immersion period.

In Fig. 2, biomass development as protein content is reported vs time.

3.1.2. Cathodic curves
Figure 3 shows the oxygen reduction curves potentiostatically obtained in the laboratory at sea water temperature close to 2°C.

The cathodic curve shifts with the exposure time as shown in the Figure and the

Table 2. Chemical composition of the sea water

Depth	Ca	F	SO$_4$	Mg	Na	K	Li	Sr	S‰	O$_2$	pH	T
m	mM	μM	mM	mM	M	mM	mM	mM	p.s.u.	%		°C
50	10.57	71.6	28.8	53	0.47	10.3	0.027	0.091	34.90	90	8.0	−1.2

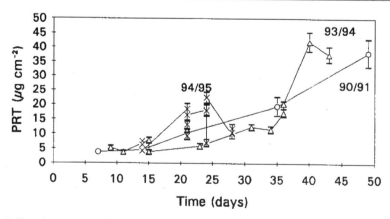

Fig 2. Protein content vs time.

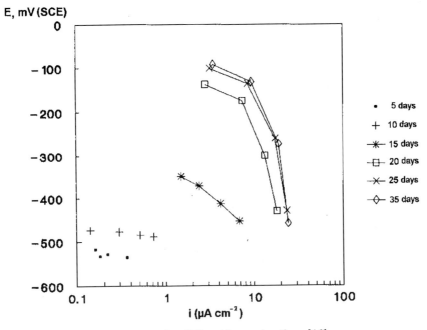

Fig. 3. Shift of cathodic curves at 2°C at different immersion times [16].

strongest change of the oxygen reduction curve occur between 15 and 25 days exposure. An extrapolation of the curve to an oxygen reduction current close to the stainless steel passivity current suggests that free corrosion potentials close to 0 mV (SCE) can be expected in Antarctica.

This level is much lower than the potential values close to +300 − +400 mV (SCE), reached in the European Seas.

3.1.3. Free corrosion potentials of stainless steels in the passive state

The trends of the open circuit potentials measured on passive stainless steels exposed at 2°C in the laboratory during the 1989/90, 1990/91, 1993/94 and 1994/95 campaigns are summarised in Fig. 4 and confirm the expected results.

It is evident that the open circuit potentials of all the stainless steels tested, never exceed the threshold values around +100 mV.

3.1.4. Corrosion potential of stainless steels in active state

During the short tests in the laboratory only AISI 304 nucleated corrosion under preformed crevice within the test period. The evolution with time of open circuit potentials reported in Fig. 5 shows that active corrosion occurs in the potential range −100 to −200 mV (SCE).

3.2. Long Term Tests

The weight losses suffered by AISI 304, 316 and 18-9 alloys, when exposed in the

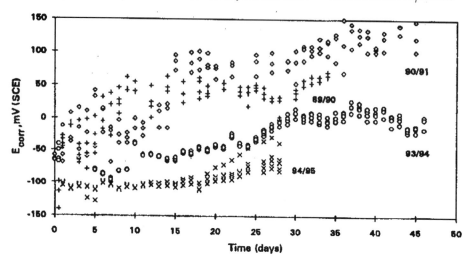

Fig 4 *Free corrosion potentials vs. time of stainless steels in passive state during the 1989/90, 1990/91, 1993/94 and 1994/95 campaigns.*

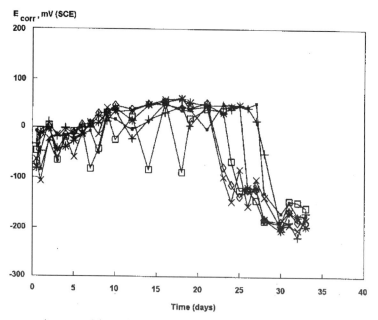

Fig. 5 *Free corrosion potentials vs time measured on six samples of AISI 304 SS type [16].*

remote crevice assembly, are reported in Fig. 6 as a function of the immersion time.

All the other stainless steels tested remained uncorroded after a 3-year exposure showing that the critical potentials for localised corrosion onset were not exceeded in the Antarctic Sea.

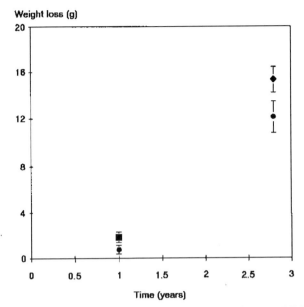

Fig. 6 Weight losses vs time measured in the remote crevice assemblies on AISI 304 (•), 316 (■) and 18-9 (♦).

4. Discussion
4.1. Biofilm Structure

The main results of the analytical approach can be summarised as follows:

– all the parameters examined (ESP, protein and carbohydrate contents) are sharply higher than those measured in the Northern European stations throughout all the year (see Fig. 7 as an example);

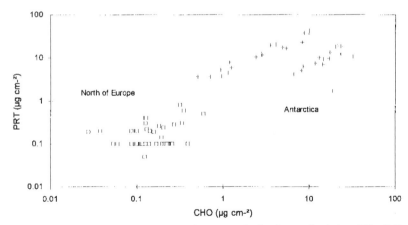

Fig. 7 Protein (PRT) vs carbohydrate (CHO) contents in the Antarctica (+) and North Europe (o) samples.

- the biological ecosystems settling on surfaces during the Antarctic summer time are dominated by algal blooms due to high solar irradiation and the presence of noticeable quantities of nutrients resulting from the melting of the ice;
- the EPS fraction, in particular, is very high compared with the values reached in the North of Europe (Fig. 8).

4.2. Electrochemical Effects of Biofilm Growth

4.2.1. Onset of localised corrosion

In spite of the high amounts of EPS, the open circuit potentials of stainless steels remained very low compared with the values reached in the European environmental conditions and, obviously, the linkage between E_{corr} and EPS amounts found in European Seas (Fig. 9) is no longer valid in the Antarctic Sea as shown in Fig. 10.

On the other hand, the fact that the potential values never exceed +50 – +100 mV SCE, when biofilm has grown on surfaces, well explains the lack of localised corrosion nucleation observed on almost all the stainless steel grades tested in the Antarctic environmental conditions.

The effects of sea water temperature on the onset of localised corrosion of stainless steels, obtained from an examination of literature data, the main results of the BIOFILM MAST II Project and the Antarctic data, are summarised in Fig. 11.

The maximum open circuit potentials vs sea water temperature reported in Fig. 11 show that a temperature decrease to 2°C and a temperature increase up to 40°C are sufficient to cancel the biofilm effect on E_{max} values.

As a consequence, sea water temperatures below 2°C and above 35–40°C abruptly reduce the risk of localised corrosion onset in natural sea water.

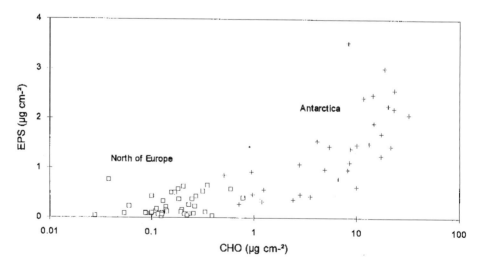

Fig. 8 *EPS vs carbohydrate contents for the Antarctica and North Europe samples.*

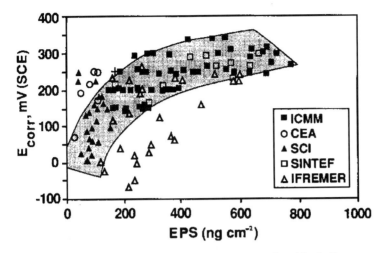

Fig. 9 *Link between free corrosion potentials and EPS contents found in the European seas [14]. Data from ICMM (Italy), CEA (France), SCI (Sweden), SINTEF (Norway) and IFREMER (France).*

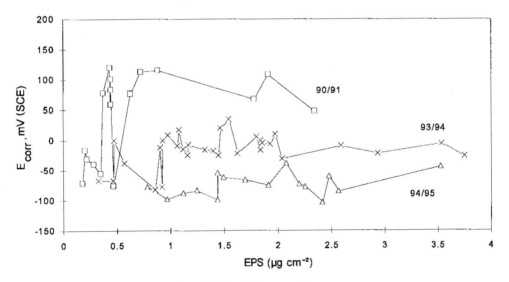

Fig. 10 *Free corrosion potentials vs EPS in all the campaigns.*

4.2.2. Localised corrosion propagation rate

As shown in Fig. 5, a stainless steel in the active state settles at a potential value close to −100 − −200 mV SCE. On the other hand, Fig. 3 showed that, in presence of a biofilm settled on surfaces, the oxygen reduction occurring at these potentials is close to the limiting diffusion currents.

Therefore, this can explain why the active dissolution of stainless steels gives rise

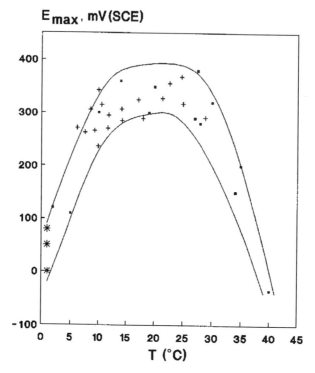

Fig. 11 *Maximum open circuit potentials of s.s. in passive state vs sea water temperatures.* (*) *Antarctic data,* (■) *ICMM data,* (+) *MAST data.*

to high corrosion propagation rates almost independent of the composition of the stainless steel and close to those occurring in the European Seas.

5. Conclusions

Biofilm growth on stainless steel surfaces leads to oxygen reduction depolarisation at the Antarctic sea water temperatures as in European sea water.

Nevertheless, the new cathodic curve is shifted toward less noble potentials compared with those observed in the European Seas.

Consequently, the Antarctic biofilms can induce only a very little ennoblement of free corrosion potentials on stainless steels in the passive state which therefore sharply reduces the risk of the onset of localised corrosion.

The biofilm presence, however, is able to sustain corrosion rates similar to those observed in the European environmental conditions.

6. Acknowledgement

The authors thank very much ENEA logistics for their great help to the group activity during the Antarctic campaigns.

References

1. A. C. Madsen, F. O. Mueller, R. Geisert, J. B. Lee, and J. Choi, *Mat. Perform.*, 1987, **26**(2), p.49.
2. R. J. Edyvean, L. A. Terry and G. B. Picken, *Int. Biodeterior. Bull.*, 1985, **21**, 4, 277–284
3. A. Mollica, A. Trevis, E. Traverso, G. Ventura, V. Scotto V. *et al.*, in *Proc. 6th Int. Congr. on Marine Corrosion and Fouling*, Athens, 5–8 Sept. 1984, 269–281.
4. V. Scotto, G. Alabiso, and G. Marcenaro, *Bioeletrochem. Bioenerg.*, 1986, **16**, 347–355.
5. R. Holte, E. Bardal and P. O. Gartland, *Corrosion '88*, Paper No.93, NACE, Houston,Tx, 1988.
6. S. C. Dexter and Y. G. Gao, *Corrosion '87*, Paper No.77, NACE, Houston, Tx, 1987.
7. A. Mollica, A. Trevis, E. Traverso, G. Ventura, G. De Carolis *et al.*, Corrosion, 1989, **45**, p. 48.
8. R. Holte, P. O. Gartland and E. Bardal, Influence of microbial slime layer on the electrochemical properties of s.s. in sea water, in *EUROCORR '87*, European Corrosion Meeting, pp. 617–623. Published by Dechema e.V., Frankfurt, 1987.
9. A. Mollica, *Int. Biodeterior. & Biodegrad.*, 1992, **29**, 213–229.
10. C. M. Scillmoller and B. D. Craig, *Mat. Perform.*, 1987, **26**, 46.
11. V. Scotto, M. Beggiato, G. Marcenaro and R. Dellepiane, Effect of microbiological and biochemical factors on marine corrosion of stainless steels, in *Marine Corrosion of Stainless Steels: chlorination and microbial effects*, European Federation of Corrosion Publications No. 10, pp. 21–33, published by The Institute of Materials, London, 1993.
12. J. P. Audouard, C. Compère, N. J. E. Dowling, D. Féron, D. Festy *et al.*, Effect of marine biofilms on high performance stainless steels exposed in European coastal water, in *Microbial Corrosion, Proc. 3rd Int. EFC Workshop*, Portugal, 1994. European Federation of Corrosion Publication No. 15, pp. 198–210, (eds A. K. Tiller and C. A. C. Sequeira), published by The Institute of Materials, 1995.
13. J. P. Audouard, C. Compère, N. J. E. Dowling, D. Féron, D. Festy *et al.*, Effect of marine biofilms on s.s. in the results from an European Exposure Program, in *Proc. Int. Congr. on Microbial Induced Corrosion*, New Orleans, May 1995. In press.
14. V. Scotto, A. Mollica, J. P. Audouard, P. Carrera, C. Compère *et al.*, Marine biofilms on stainless steels: Effect on monitoring and prevention, in *Project Reports of the 2nd MAST Days and Euromar Market*, Sorrento, Italy ,Vol. 2, pp.1059–1073, November 1995.
15. G. Alabiso, U. Montini, A. Mollica, M. Beggiato, V. Scotto *et al.*, Marine corrosion test on metal alloys in Antarctica: Preliminary Results, in *Marine Corrosion of Stainless Steels: chlorination and microbial effects*, European Federation of Corrosion Publication No. 9, pp. 36–47, published by The Institute of Materials,London, 1993.
16. G. Alabiso, V. Scotto, U. Montini, G. Marcenaro, A. Beggiato *et al.*, Biofilm interference on stainless steel corrosion behaviour in Antarctic sea water, in *Proc. 8th Int. Congr. on Marine Corrosion and Fouling*, Taranto, Italy, Published in *Oebalia*, 1993, **XIX**; Suppl., 559–566.

9

Propagation Rate of Crevice Corrosion in the MTI-2 Test for a Duplex Stainless Steel Compared to an Austenitic Stainless Steel

J. FRODIGH

R & D Centre, AB Sandvik Steel, Sweden

ABSTRACT

Selection of materials for service in sea-water and media causing microbial influenced corrosion (MIC) should be based on those materials with good resistance to the initiation and propagation of crevice corrosion. The ferric chloride MTI-2 and ASTM G48B tests are often used to determine the critical crevice corrosion temperatures. These tests have been applied to the duplex stainless steel SAF 2507 and the austenitic stainless steel UNS S31254. The conclusion reached is that the duplex stainless steel SAF 2507 experiences much lower propagation rates in the $FeCl_3$ environment compared to the austenitic stainless steel UNS S31254.

The major reason for this behaviour is believed to be (a) the higher chromium and nitrogen contents of the duplex stainless steel, (b) the presence of smaller and elongated grains in the duplex stainless steel, and (c) attack of one phase at a time for the two phase duplex stainless steel.

1. Introduction

Propagation of crevice corrosion is of importance because this type of corrosion is often of major concern in sea water and MIC environments. The selection of materials for this kind of environments should be based on those with high enough resistance to avoid initiation of crevice corrosion. However, if crevice corrosion is initiated, for example, by a temporary increase of Cl content or temperature excursion it is important to have a low propagation rate. The reason is that a deep attack is more severe and may continue to self propagate when the conditions return to normal. A shallow attack has a much greater ability to repassivate and stop growing, compared to a deep attack.

The ferric chloride test MTI-2 (or ASTM G48B) is often used to determine critical crevice corrosion temperatures (CCT) for different alloys. The critical temperature values achieved are those for which crevice corrosion starts to initiate. These values can be used to rank alloys regarding their performance in chlorinated sea water or MIC environments. In the present investigation the MTI-2 test has been conducted at a temperature above CCT and the depth and propagation rate of the developed attack studied.

The specific MIC environment consisting of iron bacteria, which are able to oxidise Fe^{2+} to Fe^{3+} and thereby create $FeCl_3$ together with Cl^- ions, as discussed by Thierry

[1], is probably simulated very well by the MTI-2 test. Almost all MIC environments also create concentration cells because of differential aeration, and this promotes crevice corrosion.

2. Background

The mechanism for crevice corrosion according to Fontana [2] is shown in Figs 1 and 2. Initially, when a crevice is exposed to an aggressive aqueous environment, the same reaction will take place both inside and outside the crevice (see Fig. 1). At the anode, metal is dissolved, and at the cathode, oxygen is reduced to hydroxide ions, see eqns (1) and (2).

Anodic $\quad M \rightarrow M^+ + e^-$ (1)

Cathodic $O_2 + 2\,H_2O + 4\,e^- \rightarrow 4\,OH^-$ (2)

After some time the oxygen in the crevice solution is consumed by the cathodic reaction. The oxygen reduction now only progresses outside the crevice and the anodic metal dissolution reaction still progresses inside the crevice (see Fig. 2). This leads to rapid attack of the metal within the crevice and no attack outside the crevice.

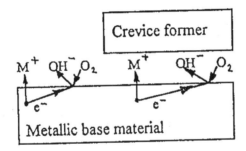

Fig. 1 Early stage of crevice corrosion.

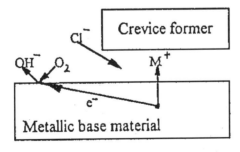

Fig. 2 Later stage of crevice corrosion.

There are some additional processes that lead to increased severity of the crevice solution. One of these is the fact that metal dissolution in the crevice creates an excess of positively charged metal ions. Therefore, there is a driving force for negatively charged ions moving into the crevice. One such negatively charged ion is Cl^-. It is well known that high concentrations of Cl^- ion initiate breakdown of the passive film.

Another process that also increases the severity of the crevice solution results from the fact that metal salts hydrolyse in water according to eqn (3).

$$M^+Cl^- + H_2O \rightarrow MOH + H^+ Cl^- \tag{3}$$

This means that the pH of the crevice solution decreases, which increases the severity.

One alternative cathode reaction during crevice corrosion is hydrogen evolution inside the crevice. This is discussed by Oldfield *et al.* [3] and Oldfield [4]. Oldfield *et al* [3] suggest that hydrogen evolution is not significant on austenitic stainless steels but is important on ferritic materials. A prerequisite of hydrogen evolution is that the pH is sufficiently low. This is likely to be the case in a crevice as discussed above.

3. Material

The materials studied in the present investigation are the duplex stainless steel SAF 2507 (UNS S32750) and the austenitic stainless steel UNS S31254. The composition of the studied materials is shown in Table 1.

All materials were in the form of cold rolled and annealed plates, approx. 4 mm thick. Test coupons were taken from the plates and ground to 120 or 80 grit paper.

The ranking of different alloys regarding crevice corrosion resistance can be determined by the pitting resistance equivalent number (PRE), calculated using the empirical formula:

$$PRE = [\% \ Cr] + 3.3 \times [\% \ Mo] + 16 \times [\% \ N]$$

With this definition and the composition according to Table 1 the PRE numbers become approximately 43 for both alloys. This means that SAF 2507 and UNS S31254 have the same temperature for crevice corrosion initiation in the ferric chloride test of approximately 37°C.

Table 1. Composition of studied materials (wt%)

Grade	C	Si	Mn	Cr	Ni	Mo	Cu	N
SAF 2507	≤ 0.030	≤ 0.8	≤ 1.2	25	7	4	—	0.3
UNS S31254	≤ 0.020	≤ 1.0	≤ 1.0	20	18	6.1	0.7	0.20

4. Experimental

The test procedures used in the present investigation were those according to the MTI-2 standard procedure and a modified version of the ASTM G48-B procedure. Both these procedures use a Teflon multiple crevice assembly with 12 crevices applied on each side of the test coupon. In the MTI-2 test a torque of 0.28 Nm is used for the fastener bolt and in the ASTM G48-B test the torque was approximately 0.3 Nm. The torque was set with a sensitive dynamo metric wrench to exactly the same value on each coupon. The torque is the most important value that must be repeated exactly to reduce the scatter in measurements of crevice corrosion. The reason for this is that the torque determines the crevice gap and a smaller gap means a decreasing circulation of crevice solution, which leads to a more severe environment within the crevice.

The test coupons with the multiple crevice assemblies applied were exposed to 6 % $FeCl_3$ solution at different temperatures. After exposure, the depth of every crevice corrosion attack was measured. This was made using an optical light microscope by focusing at the bottom of the crevice attack and at the surface of the coupon and reading the difference in μm on the focusing scale.

All tests were performed with two identical coupons from each material tested, which means that a total number of 48 (2×24) crevices were applied to each material in one test. Both grades, SAF 2507 and UNS S31254, were tested simultaneously to exclude any differences in the result as a consequence of the test procedure.

The different test parameters are summarised in Table 2.

In the first three tests two coupons were interrupted after 8 h, two coupons after 16 h and two coupons were left for the total test time of 24 h. This was done to determine the time for the initiation of the attack.

The depth measurements were chosen to be reported as the depth of the individual attack arranged in decreasing depth and also as the average depth of all the attacks versus time of propagation. From the depth vs time presentation the propagation rate was also calculated.

The pH of the solution was measured before and after the 24 h test period for test No. 4.

The microstructure and the shapes of the crevice attack were also examined in an optical light microscope.

Table 2. Conditions at the different tests

Test No.	Type	Surface (grit)	Torque (Nm)	No. of coupons	No. of crevices	Test time (h)	Test [°C] temperature
1	MTI-2	80	0.28	2	48	8	50
2	MTI-2	80	0.28	2	48	16	50
3	MTI-2	80	0.28	2	48	24	50
4	ASTM G48B	120	~0.3	2	48	24	40

5. Results

The depths of the individual attack after 8, 16 and 24 h in tests No. 1– 3 are presented in Fig. 3 for SAF 2507 and in Fig. 4 for UNS S31254. It can be seen that almost all the attacks for both SAF 2507 and UNS S31254 were initiated within 8 hours. Only a few more attacks developed between 8 and 24 h. It can also be seen that the attacks did not propagate much between 8 and 16 h. The same phenomenon is shown in Fig. 5, where the depth vs time of propagation is presented. It is obvious that the propagation rate is high in the beginning (within ≈8 h) and then decreases (between ≈8 and ≈16 h). The propagation rate then increases again (after ≈16 h) but not to the same high value as within 8 h. The propagation rate up to 8 h is probably also higher than shown here, because some incubation time before initiation of the attacks must be present.

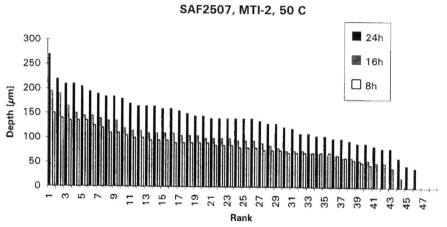

Fig. 3 Depth of crevice attack after 8, 16 and 24 h respectively for SAF 2507.

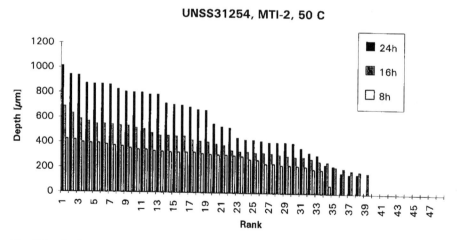

Fig. 4 Depth of crevice attack after 8, 16 and 24 h respectively for UNS S31254.

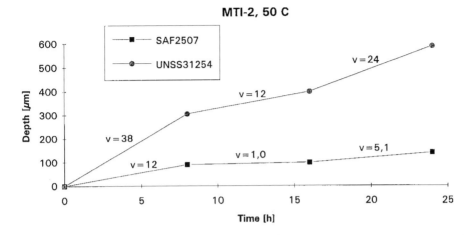

Fig. 5 *Depth vs time of propagation for both alloys; v = propagation rate in μm h⁻¹.*

In Fig. 6 the depth of the attack after 24 h is shown for the two materials. Here it is clear that the depth is larger for the austenitic stainless steel than for the duplex stainless steel. In Fig. 7 the propagation rate is shown for the two alloys based on the average depth from Fig. 6. It can be seen that this propagation rate is very similar to the propagation rate between 16 and 24 h shown earlier.

Figures 8 and 9 show the same result as above but for test No. 4 (at 40°C). The propagation rate based on the average depth on the attack developed after 24 h is slower than at 50°C.

The appearance of the microstructure and the shapes of the crevice attack are shown in Figs 10–15 (pp.114–116).

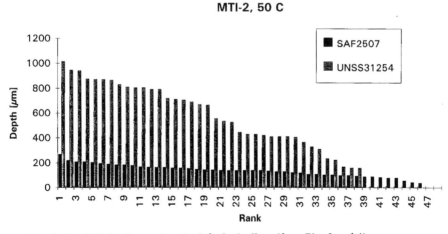

Fig. 6 *Depth after 24 h for the crevice attack for both alloys (from Figs 3 and 4).*

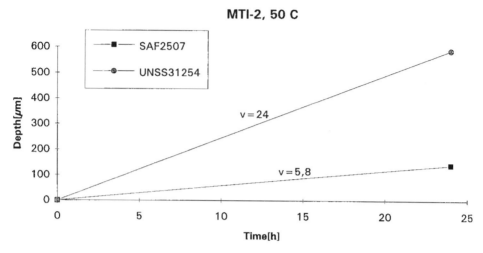

Fig. 7 *Propagation rate (v) for the two alloys based on the mean depth value in Fig. 6.*

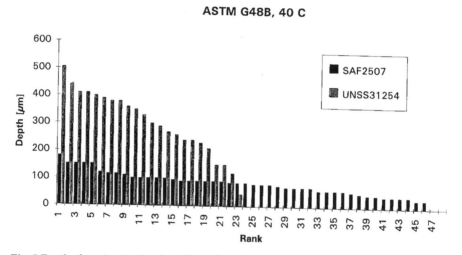

Fig. 8 *Depth of crevice attacks after 24 h for both alloys (test No. 4).*

The pH measurements of the $FeCl_3$ solution were approximately pH 1.29 before and pH 1.16 after exposure of the coupons.

6. Discussion

The relatively high propagation rate experienced before 8 h for both alloys is probably associated with the fact that the crevice in the beginning is very thin, thus preventing any dilution of the crevice solution which therefore becomes very aggressive. The

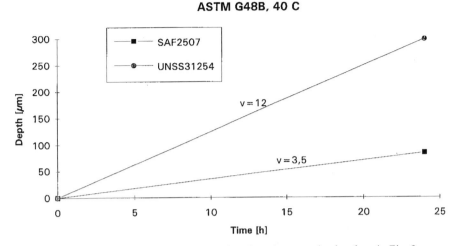

Fig. 9 *Propagation rate (v) for the two alloys based on the mean depth values in Fig. 8.*

decreasing propagation rate after 8 h should then be due to the fact that the attack in the crevice spreads and eventually reaches the crevice mouth. The solution in the crevice can then become diluted with fresh solution and become less aggressive.

The decreasing propagation rate could also be due to the fact that when the pH has decreased enough in the crevice, hydrogen evolution will start in the crevice as an alternative cathode reaction. This creates hydrogen bubbles inside the crevice, which increases the ohmic resistance and thereby decreases the propagation rate. This effect is probably more accentuated on the duplex stainless steel because the elongated and small grains trap the hydrogen bubbles better than the microstructure in the austenitic stainless steel.

The effect of hydrogen evolution is proposed by Oldfield *et al.* [3]. They also conclude that the hydrogen evolution is more likely on ferritic stainless steels, which supports the result in the present investigation.

Another reason for the lower propagation rate for the duplex steel is believed to be the alloy element content. The alloy elements that most affect the crevice corrosion resistance are Cr, Mo and N. The PRE value, which is calculated from the concentration of these elements, is the same for the two alloys, which means they have the same CCT value of approx. 37°C in the MTI-2 test. The PRE value for the duplex stainless steel is, however, based on a high Cr and N content. The PRE value for the austenitic steel is based on a high Mo content. A conclusion from this must therefore be that the PRE value reflects the resistance to initiation of crevice corrosion but not propagation rates of crevice corrosion.

Newman *et al.* [5] conclude that a high N content is generally beneficial because N reacts with H^+ in the crevice solution to form NH_4^+ ($N + 4H^+ + 3e^- \rightarrow NH_4^+$) thereby decreasing the acidity and the severity of the crevice solution. A synergistic effect between a high Cr and N contents may also be present as proposed by Osozawa *et al.* [6]. Their conclusion is that the effect of nitrogen becomes more pronounced with increasing chromium content.

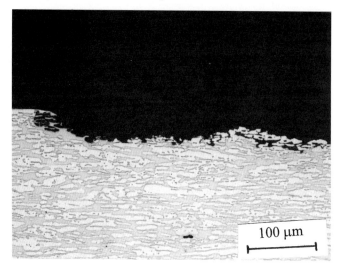

Fig. 10 *Crevice attack in SAF 2507, original surface to the left, grain size ASTM 10–12. 200×.*

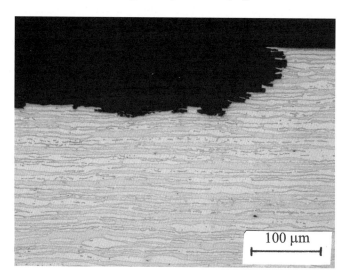

Fig. 11 *Crevice attack in SAF 2507, original surface to the right. 200×.*

With regard to the differences in the appearance of the microstructure between the two alloys, the duplex microstructure is more elongated and consists of smaller grains than the austenitic structure, see Figs 10–15. This probably means that there is a better chance to trap hydrogen bubbles — as discussed above. It is usually the case that one of the phases is attacked, i.e. sometimes the austenite and sometimes the ferrite. This is nothing unusual because the material is balanced with the same PRE value for both ferrite and austenite. The phase which is not attacked is probably locally cathodically protected by the attacked phase. This means that only about

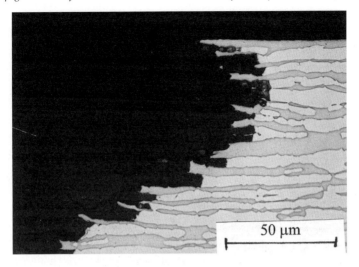

Fig. 12 *Same as Fig. 11 but 600×. Preferential attack of the austenite.*

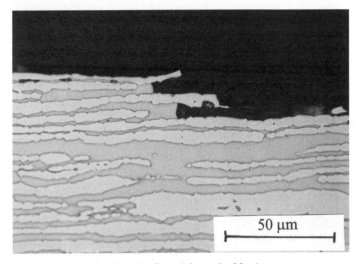

Fig. 13 *Same attack as in Fig. 11, 600×. Preferential attack of ferrite.*

50% of the material is attacked (either ferrite or austenite) resulting in low propagation rates. Which phase that is attacked is probably decided by small variations in the crevice solution. The attacked phase might also alter after some time of crevice corrosion propagation.

It must be remembered that the results in the present investigation were obtained with an $FeCl_3$ environment. For other environments the course of events for crevice corrosion propagation may be different. However, other investigations, for example those by Valen [7] have used NaCl or natural sea water and concluded that especially

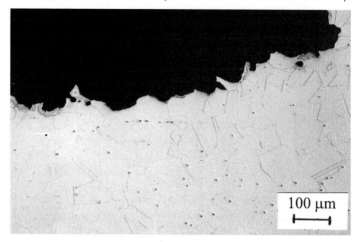

Fig. 14 *Crevice attack in UNS S312354, original surface to the right, grain size ASTM 6.5. 100×.*

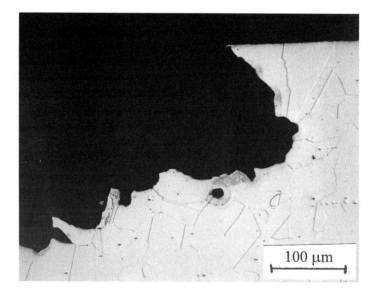

Fig. 15 *Same as Fig. 14, but 200×.*

rolled SAF 2507 achieved lower crevice corrosion propagation rates than UNS S31254. (the rolled structure has elongated grains). This suggests that the results from an FeCl$_3$ test, such as MTI-2, are relevant.

7. Conclusions

The reasons for the lower crevice corrosion propagation rate for the duplex stainless steel SAF 2507 compared to the austenitic stainless steel UNS S31254 in a FeCl$_3$ environment are believed to be:

- Higher chromium and nitrogen content in the duplex stainless steel.

- Smaller and elongated grains in the duplex stainless steel.

- Attack of one phase at the time for the two phase duplex stainless steel.

References

1. D. Thierry and W. Sand, Microbially Influenced Corrosion *in Corrosion Mechanisms in Theory and Practice* (eds P. Marcus and J. Oudar), Marcel Dekker Inc., New York, 1995.
2. M. G. Fontana, *Corrosion Engineering*, 3rd Edition, McGraw-Hill Inc., USA, 1986.
3. J. W. Oldfield, T. S. Lee and R. M. Kain, Corrosion chemistry within pits crevices, NPL, Teddington, UK, 1984.
4. J. W. Oldfield, *Int. Mater. Rev.*, 1987, **32**, 153.
5. R. C. Newman and T. Shahrabi, *Corros. Sci.*, 1987, **27**, 827.
6. K. Osozawa and N. Okato, Effect of alloying elements especially nitrogen on the initiation of pitting in stainless steel. Kawasaki Works, Nippon Yakin Kogyo Co. Ltd, Japan.
7. S. I. Valen, Initiation, propagation and repassivation of crevice corrosion of high-alloyed stainless steels in sea water. Norwegian Institute of Technology, Department of Materials and Processes, 1991.

10

Experiences with 6Mo Stainless Steel Offshore in the North Sea

R. MOLLAN

Saga Petroleum a.s.

ABSTRACT

A general review of experiences from the use of stainless steel of type 6Mo offshore on the Norwegian continental shelf is given. The review covers manufacturing aspects and corrosion resistance. Stainless steel type 6Mo has now been in use for more than 10 years on offshore platforms. The performance of the material for sea water service has been below expectations when the material was first taken into use.

The main limitation is crevice corrosion in mechanical joints, especially threaded connections. Usage limits have been defined, but are still subject to discussion between contractors/suppliers and users. There exists thorough documentation from laboratory testing and field experience, and it should now be possible to develop improved and commonly agreed usage limits.

There is a need to establish reliable, reproducible and commonly accepted standard tests for qualification of stainless steel and nickel alloys in sea water.

1. Introduction

In this paper, a general review of experiences from the use of stainless steel of type 6Mo is given.

The main application for this material has been for sea water service.

The term 6Mo stainless steel is in this paper used to cover the following alloys as defined by the UNS number:

Generic type	UNS	Nominal alloy composition (Approx.)				Commercial designations
		%Cr	%Ni	%Mo	others (%)	
6Mo	S31254	20	18	6	N=0.2	Avesta 254 SMO
	N08925	20	25	6	Cu=1,N=0.2	VDM1925 hMO
	N08926	20	25	6	N max. 0.15	UR B26
	N08367	21	24	6	N=0.2	AL-6XN

The alloys starting with "N" in the UNS number are nickel alloys, but are in this paper treated as stainless steels.

The main development of material selection for sea water piping systems for offshore projects in the Norwegian sector of the North Sea for the first 20 years can be illustrated briefly by referring to some of the larger fields as follows:

Field/Operator	Production start (Approx.)	Main material selection, main piping systems
Ekofisk/Phillips	1971	Carbon steel
Statfjord A/Statoil	1979	Cement lined carbon steel, Epoxy coated steel
Statfjord B, C/ Statoil	1982, 1985	Cement lined carbon steel, Copper nickel
Heimdal/Elf	1985	Copper nickel
Gullfaks A, B, C/ Statoil	1986, 1988, 1990	Stainless steel type 6Mo
Oseberg A, B, C/ Norsk Hydro	1988, 1988, 1991	Stainless steel type 6Mo
Snorre/Saga Petroleum	1992	Stainless steel type 6Mo

Titanium and Glass Reinforced Plastics (GRP) have been increasingly used since the middle of the 1980s and for some of the later projects this has been the dominating material for sea water systems.

For mobile production units and rigs, carbon steel has frequently been used for sea water systems, in some cases with internal hot dip galvanising.

The Snorre project was the first in which 6Mo stainless steel was used for oil and gas service. For this project the corrosion properties were thoroughly documented. It was found that the material has significantly better resistance to sulphide stress cracking than alloys such as 22Cr and 25Cr duplex stainless steels. For Snorre conditions this material has so far showed good performance in operation for hydrocarbon service.

Emphasis has been placed on what can be learned from the experiences from operation and testing, and how this so far has been reflected in the NORSOK material standards. Also, further study regarding development of standardised test methods for qualification of new materials is highlighted.

2. Manufacturing Experiences
2.1. General

The main manufacturing/fabrication limitations of 6Mo stainless steel relate to the difficulty of casting the material without cracking, the limited weldability of cast material and the importance of proper heat treatment in order to avoid precipitation of intermetallic phases — a detriment to the corrosion resistance.

On one occasion liquation/hot cracking of weld metal was found for thick walled longitudinally welded pipe. A combination of grinding and dye penetrant testing is necessary to verify this type of crack. General precautions to avoid this kind of failure are to limit the niobium content in the consumables to maximum of 0.5 % and to carry out production welding in accordance with the qualified welding procedure. An additional benefit of limiting the niobium content is to create improved fracture toughness.

On several occasions cracking has occurred as a result of contamination by copper during welding. This is not acceptable for stainless steels due to the risk of liquid metal embrittlement. This is, however, not particular to 6Mo stainless steel.

It has been found that use of Hot Isostatic Pressing, HIP, can be a cost-effective alternative for forgings.

For large components, such as valves, it has been found that carbon steel with internal weld cladding is often more cost-effective than the use of solid stainless steel.

2.2. Casting Limitations

It was realised at an early stage that the 6Mo stainless steels had the following limitations:

They are difficult to cast, especially in large dimensions, without cracking.

Welding of cast components should generally be avoided due to the tendency of cracking caused by the combination of large grains in the cast structure and the low ductility at elevated temperature.

These limitations have meant that valve and pump bodies for sea water service have to a large extent been made from other materials than 6Mo in 6Mo piping systems. At the start, copper nickel based alloys, such as nickel aluminium bronze, were often used for castings. Later, type 25Cr duplex stainless steel has been mostly used.

For cast flanges, the forging of weld necks has been carried out to improve weldability.

2.3. Sensitivity to Heat Treatment Conditions

The material properties are highly dependent on proper heat treatment and strict temperature tolerances have to be applied, especially for large wall thicknesses. It has been experienced that for cast piping components a localised corrosion rate of 6 mm in less than a year was experienced when the material was heat treated 100°C below the specified 1225°C.

In another case, the heating rate prior to forging of flanges had been too low, allowing formation of intermetallic phases. This in turn caused crack formation during the forging operation. Unfortunately these defects were not identified before most of the flanges had been installed in the piping system.

2.4. Matching Materials for Equipment

One of the main problems related to the use of highly alloyed stainless steels for sea water service is to get sea water resistant materials for all parts of equipment — such as valves and pumps that will be exposed to sea water. This problem has caused costly replacements and can probably be attributed to the following:

1. Design requirements for certain components to be of high strength, such as springs, metal to metal seals and stems. This necessitates use of alloys other than 6Mo and 25Cr duplex stainless steels; often nickel and cobalt alloys have been used, although proper documentation for performance in sea water is lacking.
2. The need for hardfacing, for which documentation of sea water corrosion resistance is scarce.
3. Lack of knowledge at contractors and suppliers, resulting in use of alloys such as nickel alloy 600 and X-750, based upon an assumption that these alloys are equal to nickel alloy 625 with respect to sea water resistance. It is often found that contractors/suppliers do not have proper knowledge of the different nickel alloys being considered for such applications.

3. Corrosion Resistance
3.1. Initial Assumption

When the material was first put into use for offshore applications some 10–15 years ago, the application limits were assumed to be as follows:

• Maximum sea water temperature: 35°C.
• Maximum concentration of residual free chlorine: 2 ppm.

3.2. Field Experience

3.2.1. General
After some time in use, crevice corrosion was experienced and it became clear that above limits had to be restricted. For those platforms in operation the level of chlorination was lowered in order to minimise the risk for crevice corrosion to the minimum required to sterilise the sea water, in the order of 0.3–0.6 ppm.

One clear observation has been that threaded connections have been more prone to crevice corrosion than other mechanical connections, such as flanges. Examples are sprinkler nozzles in fire water systems, and drainage and air venting plugs in piping systems.

One of the essential characteristics for users when comparing 6Mo stainless steel with carbon steel in sea water piping systems, is that for 6Mo the corrosion failures tend to occur within a short time after the piping systems are filled with sea water. For threaded connections such as occur in sprinkler nozzles, this can result in leakages within a few months, while for standard flanges corrosion attack is normally discovered during maintenance and/or inspection when the flanges are dismounted. Most flanges are able to tolerate quite excessive crevice corrosion before leakages occur.

It is well known that chlorination can have a detrimental effect on initiation of crevice corrosion due to its oxidising properties. On the other hand, the cathodic capacity which controls the propagation rate is reduced due to absence of the marine biofilm. Systems with and without chlorination are therefore considered separately below.

3.2.2. Systems with chlorination

Proper interpretation of field experience is difficult and where crevice corrosion has been reported for flanges, the material suppliers have often raised questions as to whether the operational conditions have been as reported.

For some fields it has been found that the outlet temperature of heat exchangers has often been higher than assumed in the design. Temperatures well above 40°C have been found, resulting in severe crevice corrosion attacks on 6Mo stainless steel. This can occur if the steady state conditions, i.e. after fouling build up in the heat exchanger, is used to determine the max. outlet temperature. In the initial period, when heat transfer is more effective, the outlet temperature can be significantly higher if the sea water flow rate is used for regulation of the main commodity temperature.

The part of sea water systems that are upstream of the heat exchangers — where the temperature is known to be in the range 5–10°C is therefore of special interest with respect to supporting the development of safe usage limits. For the Snorre platform, crevice corrosion has been reported in this part of the system, and there were no indications from control and monitoring the level of chlorination that the concentration of free residual chlorine had been above 1 ppm. The set point is now approximately 0.4–0.5 ppm.

It is important to realise that crevice corrosion is a failure mechanism which is influenced by several factors, and that these to a large extent are more or less random — one of the most important being the crevice geometry. In sea water intake conditions the probability for failure for flanged connections may, based upon the observations made so far, be as low as 0.5 to 3%. It is accordingly necessary, in a relatively large system with many crevices, to provide a large statistical basis, in order to be able to draw firm conclusions from operational experience.

Failures observed in sprinkler nozzles at and below room temperature provides a clear demonstration of usage limits. These systems are normally stagnant, and thereby the chlorine concentrations are relatively low (probably negligible for much of the time). The sprinkler nozzles are made of cast material and failure investigations have shown that the material has had a quality according to specified requirements. For sprinkler nozzles NPT threads seem to have been dominating. For Snorre the failure rate was in the order of 50% before the nozzles were replaced with new ones, and the piping system filled with potable water. Sea water will only be used in case of fire. Also, laboratory results supports the fact that the temperature limit for threaded connections is in the order of 10–20°C.

For drainage and air venting plugs forged material is normally used, and also for these components failures have been experienced in the temperature range 10–20 °C. At Snorre some 250–300 plugs had to be replaced with blind flanges.

3.2.3. Systems without chlorination

Typical applications where the sea water is not chlorinated, and where proper cathodic protection is difficult to ensure, are bolted connections and components where tolerance requirements do not allow the use of sacrificial anodes on subsea production systems.

Applications, where cathodic protection is applied, are not considered here, as properly designed cathodic protection systems will prevent crevice corrosion.

It has been a subject for discussion whether crevice corrosion of 6Mo stainless steel, and also of 25Cr duplex stainless steel, should be considered to be a limiting factor for use of these materials for subsea production systems on parts which are externally exposed to sea water at ambient temperature (5–10°C), and where cathodic protection can not be achieved. Stainless steels type 6Mo and type 25Cr duplex have recently been considered to be borderline cases and not considered as fully sea water resistant, especially for threaded connections.

Even taking into account an estimated low failure rate (in the order of 0.5 to 3%), the fact that only very few field failures have occurred is not considered to be evidence for safe use of these materials under such conditions. As pointed out above, a large statistical basis is necessary to define safe usage limits for these materials in sea water.

3.3. Laboratory Test Results and Field Trials

Extensive testing has been carried out to establish safe usage limits for 6Mo stainless steels in sea water. There is still discussion and lack of common agreement in the industry regarding what these limits should be. Nevertheless, it should now be possible, based upon all the work carried out and the available field experience, to form some conclusions on this matter.

Most tests that have been carried out have had a duration of less than 3 to 6 months, and could be used to examine conditions for initiation, and to a limited extent propagation. The understanding from previous tests had been that even though crevice corrosion has been initiated, the propagation rate would be low — although investigation of propagation rates was seldom an objective of the tests. In other words, it has previously been assumed, on not very firm basis, that the propagation rate would be low.

In 1986/87 a cast pump housing was exposed to sea water without cathodic protection at 5–18°C for a period of one year and crevice corrosion attacks with a depth of in the order of 0.5 mm were found. The pump was not running and was installed outside Bergen, on the Norwegian west coast. This clearly shows that cast material is susceptible to crevice corrosion under these conditions. Although forged material is normally considered to have slightly better corrosion resistance than cast material, this illustrates that the properties are marginal — even for conditions without chlorination. During later testings new attacks occurred, but with somewhat lower corrosion rates.

Based upon available test results it has been proposed that the maximum tolerance temperature with chlorination should be in the order of 5°C lower than the limit without chlorination.

3.4. Quality Control Testing

For the first projects, the requirement for purchase of piping bulk materials was ASTM G 48 Method A, with the pitting test carried out in 6% ferric chloride solution, and an acceptance criterion of 35°C, combined with a requirement of the material being free from intermetallic precipitates when etched and micrographed under certain conditions.

Later it was found that quantitative micrography was difficult to use both in practice and in contractual relations, only corrosion testing is now used. It has been realised that testing 6Mo materials at 35°C is not adequate enough to identify material that has substandard quality. In order to differentiate between material which has the necessary corrosion resistance, and material of substandard quality, the test temperature has been raised.

Below, the test temperatures currently required in relevant NORSOK material standards are given, and for comparison the requirements for 25Cr duplex stainless steel has also been given. Additional acceptance criteria include no pitting at 20 × magnification and a maximum weight loss of 4.0 g m^{-2}.

Acceptance criteria for temperature given in NORSOK M-CR-601/630 for ASTM G 48, Method A tests:

Type of product	Material	
	6Mo	25Cr duplex
Bulk piping components, including welded fittings	50°C	50°C
Installation welds	40°C	35°C

Pitting tests are used although the limiting factor for 6Mo stainless steels is the crevice corrosion resistance. Crevice corrosion tests are not sufficiently reproducible to be suitable for quality control tests.

4. Usage Limits
4.1. Current Limitations

The limits now widely used for 6Mo stainless steels for offshore projects in Norway are given in NORSOK standard M-DP-001 "Material Selection". These were developed in the autumn of 1994 and the recommendations given may be summarised as follows:

1. Sea water systems with crevices:
   Max. operating temp. 15°C, max. residual chlorine 1.5 ppm.

2. Sea water systems without crevices:
   Max. operating temp. 30 °C and same max. chlorine level.

3. For subsea applications at ambient conditions without cathodic protection 6Mo and 25Cr duplex stainless steels are considered borderline cases and not fully sea water resistant.

4. These materials should not be used for threaded connections without cathodic protection for subsea conditions.

5. Type 6Mo stainless steel can be used in sea water systems with crevices above 15°C if crevices are weld overlayed. Where weld overlay is used to prevent crevice corrosion in sea water systems, alloys with documented crevice corrosion resistance in the as-weld overlayed condition shall be used. The maximum temperature shall be documented, ref. item 6.
    (As accounted for below, nickel alloy 625 is not a proper selection. The relation between chemical composition and corrosion resistance is not well understood, but it seems that Alloy 59 is a possible candidate.)

6. Recommended test for qualification of weld deposits for weld overlays with respect to crevice corrosion resistance:
    The MTI test procedure, MTI Manual No. 3, using a tightening torque of 2 Nm. (The selected tightening torque has been established based upon recent results, and correlation to operational conditions should be verified.)

These limits have been considered to be too strict by several contractors and material suppliers and the limits will be reconsidered in connection with the revision process which have been ongoing during the first half of 1996.

There are indications that even the max. temperature obtained in the modified MTI test, i.e. with tightening torque of 2 Nm, is too high, and that the acceptance criterion should therefore be set 10–20°C higher than the max. usage temperature.

The above limitations have been considered to be too strict by several contractors and material suppliers, and the limits will be reconsidered in connection with the revision process, which will be ongoing during the first half of 1996. Currently the following philosophy can be considered justified for subsea sea water exposure where cathodic protection cannot be achieved:

(a) Stainless steel type 6Mo and 25Cr duplex are prone to crevice corrosion in sea water, even at temperatures as low as 5–10 °C, but the probability for failure is low.

(b) Accordingly it is recommended to use these materials with caution for critical components, even at ambient temperature. These materials can be used for non-critical systems and systems with redundancy.

(c) Threaded connections, such as for bolts, are considerably more prone to crevice corrosion than regular planar mechanical connections, such as flanges.

It is important to realise that nickel alloy type 625 is only slightly better than stainless steel type 6Mo, with a critical crevice corrosion temperature in the order of 5°C higher than 6Mo for forged materials. In the as-cast condition, including as-overlayed, the crevice corrosion resistance of nickel alloy type 625 is less than for forged stainless steel type 6Mo. Nickel alloy 625 is therefore not able to improve the crevice corrosion properties of 6Mo stainless steel.

4.2. Design Issues

Based upon results from laboratory testing it seems to be a reasonable approach not to use graphite-type gaskets for 6Mo stainless steels, but to employ aramide or glass fibre reinforced rubber gaskets.

Laboratory testing has indicated that it is advantageous initially, for a period of approximately 2 weeks, to use the piping systems without chlorination and at ambient temperature, thereby building up a slightly better resistance towards crevice corrosion.

The practical consequences of not using threaded connections in a piping system are more dramatic than first appear.

In normal piping design it is common to use drainage and air venting plugs that are screwed into fittings welded to the main pipe. These can, if required, be substituted by blind flanges.

In many standard types of valves used for sea water service the housing contains drainage plugs screwed into the body.

Traditional firewater systems have valves that are screwed directly into the piping.

If threaded connections cannot be allowed due to the limited crevice corrosion resistance, this will necessitate costly changes implying use of special deliveries for piping and equipment.

5. Development Needs
5.1. Usage Limits

In accordance with the above further work is needed to improve the definition of the usage limits for 6Mo materials. The following limits need to be covered: A temperature limit for piping systems with chlorination, for flanged and threaded connections, respectively and with limitation on chlorine concentration.

As above, but without chlorination.

Temperature limit for subsea applications without cathodic protection, for flanged and threaded connections, respectively.

5.2. Qualification Tests

Tests to document performance of new materials in sea water service are needed. Such tests should be reliable, reproducible and commonly accepted, and should prior to standardisation preferably be subjected to a "round robin" test programme to verify consistency between laboratories.

5.3. Material documentation

From the design requirements there is a need for the use of materials other than 6Mo and 25Cr duplex stainless steels in sea water systems, e.g. high strength materials for components and hardfacing. There is a need to provide better documentation for nickel alloys and cobalt alloys as well as hardfacing materials in this respect.

6. Conclusions

• Stainless steel type 6Mo has now been in use for more than 10 years on offshore platforms. The performance of the material for sea water service has been below the expectations when the material was first taken into use.

The main limitation is crevice corrosion in mechanical connections. Of these, threaded connections are more prone to crevice corrosion than flanges and other more or less planar mechanical connections.

• During the last 10 years extensive laboratory testing has been carried out and it should therefore now, based upon results from laboratory testing and field experience, be possible to establish safe usage limits for this material.

The fact that the usage limits for 6Mo stainless steel up to now has been subject to disagreement between contractors/suppliers and users, even though all this experience is available, illustrates the difficulty in developing suitable tests for qualification of stainless steel and nickel alloys for use in sea water. Further the difficulties involved in proper understanding of field experience.

• There is a need to establish reliable, reproducible and commonly accepted tests for qualification of stainless steel and nickel alloys in sea water. The test procedures should be subject to 'round robin' testing prior to standardisation to verify consistency between laboratories.

11

An Innovative Low Cost Solution Combating Corrosion Problems in the Draugen Sea Water System

I. H. HOLLEN

A/S Norske Shell - Draugen Operations, Norway

ABSTRACT

The 6Mo based sea water utility system on the Draugen platform formed leaks just after start up due to the high temperature at the sea water outlets from the process coolers. Several solutions were assessed and a concept from Sintef named "Resistor-controlled Cathodic Protection — RCP" was selected, and developed into a product that was successfully installed in a joint co-operation with CorrOcean A/S. The innovative RCP concept cost approx. 10% of the other solutions evaluated such as material changeout. The RCP concept function was based on results from the first months in operation and it is being evaluated by other operators to solve similar problems as those on the Draugen field. It can furthermore provide an attractive solution when combined with cheaper, less corrosion resistant alloys (e.g. 316L), giving better life cycle costs than the more costly steels commonly used today.

1. Introduction

The Draugen field is located on Haltenbanken and came on stream in the autumn of 1993. The concept is based upon an unique monotower concrete Gravity Base Structure situated in some 250 m of water depth, supporting an integrated platform deck with production, drilling and living quarter facilities. The field development further comprises a floating loading buoy, subsea satellites for production, water injection and gas injection. Draugen is an oil field with an average production of approx. 150 000 bbl/d at present. This paper outlines the major material related problems experienced with the Draugen sea water system, and focusses on how it was solved by industrialising a new concept of protection within a limited time scale. [See also Drugli *et al.*, this volume, p.165, and Rogne *et al.*, p.69.]

2. The Draugen Sea Water System

The installed sea water systems on the Draugen platform are manufactured from austenitic stainless steel with 6% molybdenum; the so called '6Mo' stainless steel. A design limit of 35°C was set for the sea water temperature since 6Mo is vulnerable to crevice corrosion at higher temperatures. The choice of 6Mo, which is rather expensive, was based on having a maintenance-free material giving favourable life cycle costs. After start-up the temperatures at the sea water outlets from the coolers were registered as high as 67°C and it was clear that corrosion would be experienced and a search for the best solution started. The first leak soon occurred and more followed — in welds and crevices at flanged connections.

3. Searching for Solutions

Draugen is designed using a 'single train – no sparing' approach, leaving little opportunity for bypass or part system shutdown. The sea water based cooling system is essential for steady production and hence could not be allowed to break down. Since time was essential it was necessary that system intervention/repair should be planned for the annual scheduled shutdown to minimise production loss. This meant managing the leaks that were discovered and those expected until the repair could take place. Furthermore, the time allowed for finding, preparing and implementing a good repair solution was limited to five months. Several possible solutions were assessed as follows:

- Installing a sea water bypass feeding cold sea water to the cooler outlet in order to reduce the damaging high temperature. This solution was not feasible on all locations due to lack of sufficient cooling capacity, and the work would also be costly and require shutdown.

- Another feasible but costly option found was the replacement of the 6Mo steel with titanium. The long lead time (approx. 6 months) for titanium was also a concern and finally a full material change would require a considerable plant shutdown time.

- Using titanium on critical components in combination with weld overlay with high alloyed Ni-alloys (e.g. UNS 10276) was an alternative to full titanium replacement. This, however, was still costly and dependant on a shutdown because of the complexity of this operation offshore.

- Glassfibre Reinforced Epoxy — GRE — was assessed as a replacement material, but the process conditions gave concern with regard to its suitability. Furthermore, the mechanical parts ,such as valves, would have to be replaced with higher grade steel.

- Lastly, a concept from Sintef and CorrOcean A/S, called the 'Resistor-controlled **C**athodic **P**rotection — RCP' was proposed, using cathodic protection to protect stainless steels. This concept was simple; it was possible to industrialise in a short time and its robustness meant that there would be very little follow up and maintenance costs over the field life. Finally, it was approximately 10–15% of the costs of the other solutions evaluated and could be fitted within the scope of an already planned shutdown. The cost of the RCP concept — including design, fabrication and installation was approx. $100 000 covering 10 anode locations.

4. The 'Resistor-controlled Cathodic Protection – RCP' concept

The basic principle of the concept developed by Sintef is to apply cathodic protection to a pipe system of stainless steel using a resistor 'R'in series with the anode 'A' to control both the potential on the stainless steel and the anode current output 'Ia'. The principle is shown schematically in Fig. 1.

The RCP method is further based on the observation that the protection potential for prevention of local corrosion on stainless steels is much more positive than the typical potentials of sacrificial anodes. The voltage drop over the resistor is therefore designed to obtain sufficient, but not excessive, negative polarisation of the stainless steel. The resistor control thus keeps the stainless steel in a protective potential range, where the current requirements are very small in many saline environments — e.g. in chlorinated sea water, in produced water and in natural sea water above 30°C. Due to very low current requirements in the relevant potential range, a single anode can protect large lengths of the pipe system at a very low anode consumption rate.

5. From Concept to Product

Due to the time constraints in relation to the next planned annual shutdown, only 4–5 months were available to take the solution from a concept to a useful and reliable product. A co-operation was therefore initiated between Sintef and CorrOcean A/S. The latter are Norske Shell Draugen Operations partner on corrosion and material matters, and have good experience of industrialising products. The CorrOcean standard access fitting was chosen as the basis for the design of the anodes, hence allowing the use of a standard hydraulic retriever tool for anode installation and future replacement. Full laboratory trials were proposed to demonstrate the concept — especially since repassivation of already attacked areas was a concern. However,

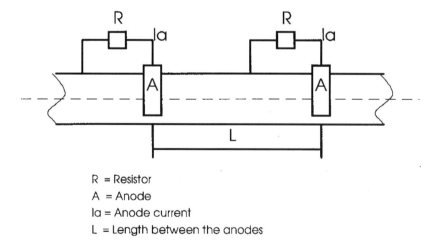

R = Resistor
A = Anode
Ia = Anode current
L = Length between the anodes

Fig. 1 *Schematic of the RCP method applied to a piping system.*

it was decided that, in order to save time, some conservatism in the design should be applied and followed up when the system was installed offshore to gain the required experience and feedback.

The system was successfully developed and installed during the January 1995 shutdown. Installation involved welding on the access fittings at the correct locations on the pipework and then inserting the anodes. The anodes were located on the sea water line just after the cooler, at a distance of 2–3 m away. One anode was deemed sufficient per cooler outlet line, since it protected the pipe length required for the water temperature to drop below the critical area. In practical terms on Draugen this means the water enters a manifold and is cooled down. A regular monitoring of the system is now implemented, using a portable instrument to collect the data. The field data are important here because they fulfil both the need for monitoring the system function and also act as a real field trial and source of feedback to the Sintef and CorrOcean staff to improve and optimise future designs. It is already clear that, when the one year verification and data collection period is ended, there will be very little need for system maintenance and follow up — giving favourable life cycle costs and supporting the investment already made in 6Mo.

The design of a crude cooler modification on Draugen involving some new piping coincided with the RCP project. Since it involved 6Mo piping already installed, the RCP concept was incorporated in the modification design as the best overall solution in this case. Thus, it must be accepted that even a stainless steel may need some help to survive, and that RCP is a design as well as a repair option that should be used giving the most optimum solution.

Initial readings from the system indicate that it performs as intended, and the product is now being installed on other platforms to solve similar problems as on Draugen. For A/S Norske Shell, it is very encouraging to note that this concept has kept in line with the aspirations of our support strategy, comprising the use and development of local firms with a fresh view on the challenges which face offshore operations.

6. Conclusions

- The 6Mo stainless steel used on Draugen failed primarily because the process conditions were outside the original design; a water temperature that was too high led to the occurence of crevice corrosion.
- Among the solutions assessed, the concept of RCP was found to be the simplest and most economic solution, and has been successfully developed as a reliable product by Sintef and CorrOcean A/S.
- With the RCP method, the critical limits of temperature, chlorine and oxygen levels can be raised considerably without any risk of corrosion. This provides increased flexibility in the operation of the system without affecting technical integrity on both new and existing systems.
- The concept also enables the use of less corrosion-resistant alloys, with large savings on material cost.

- The initial feedback from the system onboard Draugen shows that it performs as calculated, and solves a serious corrosion/operational problem at low cost while supporting the investment already made in 6Mo.

12

Experiences with High Alloyed Duplex Stainless Steels in Sea Water Systems

R. FRANCIS, G. BYRNE and G. WARBURTON

Weir Materials Ltd, Park Works, Newton Heath, Manchester M40 2BA, UK

ABSTRACT

The paper describes the origin, composition and properties of the super duplex stainless steels. Zeron 100 was the first of the super duplex stainless steels, and some service experiences in the marine environment are described. The experiences have been good with very few problems. Some more novel applications for super duplex stainless steel in the marine environment are described.

1. Introduction

The super duplex stainless steels were first developed as cast alloys in the early 1980s, about the same time as the 6Mo austenitic alloys were being introduced. The combination of corrosion resistance and high strength made these alloys very successful for sea water, firewater and injection pumps. The success of the cast alloy led to the development of a wrought equivalent for piping, fittings, flanges etc. The first of these, Zeron 100 (UNS S32760) was introduced in 1986, and the first major sea water system was ordered the following year. Since that time super duplex stainless steels have been used extensively for sea water systems, oil and gas processing systems, and in chemical, flue gas desulphurisation and many other environments.

The super duplex stainless steels all have a ferritic/austenitic microstructure in a roughly 50:50 ratio. The high alloy content means that the alloys combine good corrosion resistance, particularly in chloride containing environments, with high strength, enabling the use of thin wall sections in high pressure/temperature environments. The alloys also have good ductility and weldability. Although there are a number of super duplex alloys available, they are all proprietary alloys, and the compositions of the most common ones are shown in Table 1. It can be seen that the nominal composition is 25 Cr–7Ni–3.5 Mo–0.25 N, and some of the alloys have additions of copper and/or tungsten as well.

The high strength of super duplex stainless steel is indicated by the minimum mechanical properties in Table 2. These are compared with 316L and 6 Mo austenitic alloys, and the greater strength of super duplex alloys, particularly the 0.2% proof stress, can be seen.

These alloys were originally developed for sea water service and they have been in use for pumps since 1986 and as pipe and fittings since 1989. Zeron 100 was the

Table 1. *Nominal composition of some super duplex stainless steels*

UNS No.	Common Name	Composition (Wt%)						PREN*
		Cr	Ni	Mo	N	Cu	W	
CAST J93380	Zeron 100	25	8	3.5	0.25	0.7	0.7	> 40
CAST J93404	2507	25	7	4	0.28	—	—	42
WROUGHT S32760	Zeron 100	25	7	3.5	0.25	0.7	0.7	> 40
WROUGHT S32750	2507	25	7	4	0.28	—	—	42
WROUGHT —	52N+	25	7	4	0.28	1.5	—	42
WROUGHT S32740	DP3W	25	7	3	0.28	—	2	39

* PREN = %Cr + 3.3% Mo + 16%N

Table 2. *Minimum mechanical properties for some stainless steels*

Alloy	0.2% Proof Stress (MPa)	UTS (MPa)	Elongn. (%)	Hardness (HRC)
316L	210	500	45	22 max.
6Mo Aust.	300	650	35	22 max.
Super Duplex	550	750	25	28 max.

first of the super duplex stainless steels and some examples of its use in pumps are given in Table 3, and as pipe and fittings in Table 4. Super duplex alloys can also be readily drawn down to heat exchanger tube sizes and many kilometres are in service in both sea water cooled heat exchangers and in subsea umbilicals.

The aim of the present paper is to describe the service experiences with super duplex and these will principally be with the author's company's product, Zeron 100.

2. Service Experience

Zeron 100 pumps have been operating in sea water since 1986 and Zeron 100 piping systems since 1989. Super duplex stainless steels have been used since the late 1980s instead of carbon steel or copper nickel because of their improved corrosion resistance, particularly at high velocities (> 3 m s^{-1}). The utilisation of smaller diameter piping led to topsides weight savings on offshore platforms compared with carbon steel or copper nickel. Over 30 systems have now been installed in the North Sea and the Far and Middle East. There have been a few problems, usually at start up or soon after, but largely the material has worked well all over the world.

Table 3. Installation list for sea water pumps in Zeron 100 (up to 1990)

Application	No.	Type	Delivery Date
Shell	1	vertical in-line	1986
Shell Oman	2	booster pump	1987
Amerada Hess Ivanhoe/RobRoy	7	lineshaft pumps	1987
Mobil Statfjord 'C'	1	injection	1988
Shell Kittiwake	2	injection	1988
Shell Gabon	2	injection	1988
Shell Osprey	2	injection	1988
Global/RDS, South Morecambe	2	3 stage submersible	1989
Shell Brent Charlie	3	2-stage tank pumps	1989
Shell Draugen	2	injection	1990
Statoil Gulfax 'B'	2	injection	1990
Home Oil Swan Hill	2	injection	1990

Table 4. Major sea water and firewater systems in Zeron 100

Client	Contractor	Project
Amerada Hess	Brown and Root Vickers	Ivanhoe/Robroy
Mobil		Beryl A Firewater system
Mobil	Davy McKee	Beryl A Deluge system
British Gas	Global	Morecambe Bay Sea water discharge
Mobil	Wimpey	Beryl A Gas sales riser
Woodside	Davy McKee	Goodwin A Sea water/firewater
BP	Brown and Root Vickers	BP Bruce
Amerada Hess	Foster Wheeler	Scott Sea water system
QGPC	Global Eng	Diyab Hook up system
Hamilton Oil	Brown and Root	Liverpool Bay sea water/firewater

2.1. Pumps

Cast Zeron 100 has been used for sea water pumps since 1986. Super duplex sea water pumps and firewater pumps have been used extensively in the North Sea and also elsewhere around the world, at ambient temperatures from 0 to 40°C. These pumps are usually supplied with schedule 10S column pipes in wrought Zeron 100. The high strength of the alloy means that thicker wall piping is not usually required to support the weight of the pump. Zeron 100 bolting is usually supplied for joining the column pipes. There have been no reports of corrosion problems with any of these pump systems.

Super duplex stainless steels are very attractive for sea water injection pumps. Current practice is to deaerate the sea water prior to injection, and so corrosion resistance is not a primary requirement. The high strength of super duplex stainless steel makes it very attractive because of the reduced wall thicknesses possible compared with austenitic alloys. Weir Pumps have supplied injection pumps operating up to 380 bar. Barrel casing pumps are traditionally cast, but hot isostatic processing (HIP) has been used for the barrel. HIP material has the higher strength of the wrought material and thus permits further wall thickness savings.

New developments could lead to the injection of raw sea water. Super duplex stainless steel pumps will be essential then because of their combination of high strength and corrosion resistance.

With super duplex pumps it was common, in their early development, to use a nickel alloy, K-500 (UNS N05500), for pump shafts. This was believed to have good resistance to sea water. Weir Pumps experienced a shaft failure where pitting had occurred at a seal and this acted as a stress raiser — leading to a fatigue failure. As the water was polluted this was not too surprising, as K-500 has poor resistance to sulfides. However, a second failure occurred in clean sea water, which appeared to have started as crevice corrosion on the K-500 at a seal. Subsequent exposure tests of K-500 to K-500 crevices and K-500 to Zeron 100 crevices showed that K-500 is susceptible to crevice corrosion in sea water, and when it is coupled to Zeron 100 the attack becomes much worse (Table 5). The alloy is clearly not compatible with high alloy stainless steels in aerated sea water, and Weir Pumps now use super duplex stainless steel shafts.

2.2. Galvanic Corrosion

The Amerada Hess Ivanhoe/Rob Roy platform reported leaks at some flange joints in the sea water system soon after start up. Investigation revealed that the gaskets used in these joints contained substantial quantities of graphite, which is well known to be highly cathodic to most metals. Hence the localised corrosion of the flanges was not surprising. The flange faces were cleaned up and re-assembled using non-graphitic gaskets. There have been no further leaks since that time.

Table 5. Maximum depth of attack for galvanically coupled crevice specimens

Couple	Alloy	Maximum pit depth (mm)	
		washer/plate	plate/plate
1	K-500	0.26	0.33
	K-500	0.12	0.30
2	Z100	0.00	0.00
	Z100	0.00	0.00
3	Z100	0.00	0.00
	K-500	0.32	0.35
4	Z100	0.00	0.00
	K-500	0.42	0.44
5	Z100	0.00	0.00
	K-500	0.52	0.50

2.3. Corrosion of Welds

Soon after start up the Amerada Hess platform, Ivanhoe/Rob Roy reported a leak at a weld in a Zeron 100 sea water pipe. However, on subsequent examination it was shown that the weld had been made with 316L stainless steel consumables. The repair, with Zeron 100X consumables, is still in service.

Zeron 100 piping has been used for both the sea water and firewater systems on the Woodside Goodwyn "A" Platform. Soon after start up a series of leaks appeared, associated with welds, in the firewater system. Weir Materials was involved in the investigation which was carried out in Australia. All of the welds were very dark on the root side, and many were covered with thick, black oxide, as shown in Fig. 1. The corrosion occurred at the low temperature heat affected zone as extensive pitting. The microsections showed extensive sigma phase precipitation in the HAZ, up to ~20%. The samples with dark films clearly indicate the use of excessive heat inputs, above those specified in the weld procedure. It is well documented that the use of high heat inputs results in the precipitation of sigma phase. The welds with heavy coking on the root side suggest little or no argon purging occurred. Under these circumstances the weld pool becomes very sticky and greater heat input is necessary to complete the weld. As described previously, excessive heat inputs in schedule 10S pipes (2" to 6") are well known to result in sigma phase precipitation.

There was little radiography carried out on the welds compared with North Sea practice. In addition parts of the flooded firewater piping were exposed at temperatures up to, and possibly exceeding 50°C.

2.4. Fatigue

One problem which was experienced by Amerada Hess was a fatigue failure of schedule 10S pipe adjacent to a large valve. The valve was a heavy 900 lb valve which had been welded into the line without adequate support, and vibration had

Fig. 1 Appearance of weld root removed from Goodwyn "A" Platform (NPS 6 schedule 10S) ×3.

resulted in fatigue and failure of the pipe. This is essentially a design problem during installation.

There have been a number of problems associated with fatigue failure of thin walled piping on North Sea platforms. These have involved not only super duplex stainless steel, but also 6Mo austenitic and 22 Cr duplex alloys. All the failures have occurred at small diameter pipe take-offs from large diameter schedule 10S pipes. The branch is attached by a weldolet and has not been adequately supported so that vibration from a nearby pump or a partly throttled valve leads to fatigue. The problem arises because one size of weldolet is used for all pipe sizes from schedule 10S to schedule XXS. This creates a notch at the toe of the weld which is the initiation site for a fatigue crack (Fig. 2) because of the stress concentration. The solution is to provide proper support for the take off pipework, and to use either a light-weight weldolet or a reducing tee. Figure 3 compares the fatigue data for 22 Cr duplex and 25 Cr super duplex. This shows the high resistance of super duplex stainless steel to fatigue.

2.5. Chlorination

Chlorine is added to sea water to control fouling. However, it is well known that chlorine increases the susceptibility of stainless steels to crevice corrosion [1–3]. Shortly after start up several leaks were reported on the sea water discharge piping

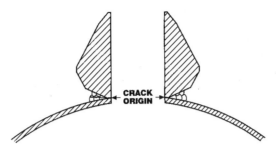

Fig. 2 Schematic diagram of site of fatigue cracks at weldolets on thin walled pipe.

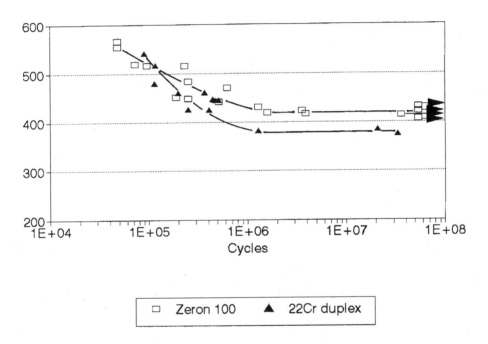

Fig. 3 S–N curves for duplex stainless steel in rotating bend tests at 50 Hz in air.

on the Scott platform. All the leaks were associated with welds, but it is not known whether the corrosion was associated with the HAZ or the fusion line of the weld metal. The sea water temperature was 42°C and the chlorine concentration was 2 mg L^{-1}. This chlorine level was obviously too high at the concomitant discharge temperature. The chlorine concentration was reduced to 0.8 mg L^{-1} at the inlet manifold and the sea water temperature to 36°C. Since the changes there have been no further failures.

The fact that no failures occurred on the inlet piping at ambient temperature shows the strong effect of temperature on the reaction between chlorine and super duplex stainless steel in sea water.

2.6. Temperature

There have been a number of problems of leakage due to crevice corrosion associated with 6 Mo austenitic alloys [4, 5]. Many of these are associated with high temperature excursions, to 60–70°C, usually for a short period. There have been no reports of problems with super duplex under similar service. On the contrary, experience at high sea water temperatures has been good. The discharge lines on Piper Bravo ran at 20°C for two or three months after start up, but as more wells were brought on stream the temperature on three lines from gas coolers rose to 60°C and stablised at 52–55°C. All three lines ran for two years with no leakage. To increase production efficiency the discharge temperature has now been increased to 65°C, and there have been no reported problems after six months operation.

Laboratory tests have been carried out to compare the resistance of 6Mo austenitic and Zeron 100 to crevice corrosion in sea water under temperature upsets. The testing has been described in detail elsewhere [6]. All of the 6Mo specimens suffered attack soon after the temperature increased to 70°C and on cooling, repassivation occurred at 30–35°C. The crevice corrosion was severe and extensive. Only half of the Zeron 100 samples broke down at 70°C and all of these repassivated at temperatures from 42 to 65°C on cooling (Table 6). These results are in agreement with the service experience and demonstrate the resistance of Zeron 100 to crevice corrosion in elevated temperature sea water.

Table 6. Repassivation data for stainless steels during the cooling cycle

Alloy	Specimen No.	Repassivation Temp. (°C)	Time to repassivation (h)
Zeron 100	A	No pitting	0.0
	B	42	6.6
	C	No pitting	0.0
	D	65	1.0
	E	48	5.7
	F	No pitting	0.0
	G	58	2.9
	H	No pitting	0.0
	I	No pitting	0.0
	J	No pitting	0.0
	K	55	4.1
6Mo Austenitic	1	37.3	15.1
	2	*	*
	3	33.5	13.0
	4	25.0	30.0
	5	47.0	8.4
	6	30.0	21.9
	7	30.0	21.2

* "breakout" occurred; see Ref. [6].

3. New Applications

In addition to the conventional uses of Zeron 100 in sea water some more novel applications have recently been installed. Corrosion of electrical conduit on offshore platforms where sea water splashing can occur is common and Zeron 100 pipe has been used for conduit on several North Sea platforms to prevent this problem.

The combination of high strength and resistance to sea water has been useful for some proprietary marine fixings into concrete. These fixings require no loss in strength over the 20 year design life. The experience to date has been excellent.

Another problem is with corrosion of steel reinforcing bar in concrete in marine environments. Not only is the marine environment corrosive to rebar steel, but it is not uncommon for sea water to be used in mixing the concrete in some parts of the world. Weir Materials was asked to supply re-bar in Zeron 100 for concrete in this environment. This is a novel extension to the applications of super duplex stainless steel in marine environments.

4. Conclusions

1. Super duplex stainless steels have excellent resistance to crevice corrosion in sea water even at elevated temperatures.

2. Service experience has been good with very few problems. Most of these problems are associated with poor design, or misuse of the material.

References

1. E. B. Shone, R. E. Malpas and P. Gallagher, *Trans. Inst. Mar. Eng.*, 1988, **100**, 193.
2. B. Walleen and S. Henriksson, *Corrosion '88*, Paper 403, NACE, Houston, Tx, March 1988.
3. P. Gallagher, A. Nieuwhof and R. J. M. Tausk, Marine Corrosion of Stainless Steels: Chlorination and Microbiological Effects, p. 73. EFC Publication No. 10, Published by The Institute of Materials, London, 1993.
4. R. Johnsen and S. Olsen, *Corrosion '92*, Paper 397, NACE, Houston, Tx, 1992.
5. R. E. Lye, Engineering Solutions to Industrial Corrosion Problems, 1993, Paper No. 2, Sandefjord, Norway, June 1993. NACE / NITO.
6. R. Francis, J. Irwin and G. Byrne, *Duplex Stainless Steels '94* , 1994, Paper 22. Glasgow, Scotland, November 1994. Published by TWI.

13

New Generation High Alloyed Austenitic Stainless Steel for Sea Water Systems — 654 SMO (UNS S32654)

B. WALLÉN

Avesta Sheffield AB, S-774 80 AVESTA, Sweden

ABSTRACT

Tests in hot chlorinated sea waters have been performed at two sea water laboratories to compare the corrosion resistance of the 7Mo superaustenitic stainless steel 654 SMO®* with that of the 6Mo stainless steel 254 SMO®* and the nickel base alloys 625 and C-276. 654 SMO was much more resistant to crevice and pitting corrosion than 254 SMO and at least as resistant as the best nickel base alloy. Continuous chlorination results in appreciably more corrosive conditions than intermittent chlorination. If the continuous chlorination is preceded by a period with intermittent chlorination, the corrosion resistance of a stainless steel will increase.

1. Introduction

Since the beginning of the 1980s, great amounts of the superaustenitic 6Mo stainless steel Avesta Shefffield 254 SMO are being used in the sea water systems on North Sea offshore plafforms. Although the experience generally has been very good [1, 2], some cases of crevice corrosion have recently been reported [3–5]. These attacks have mainly occurred in threaded connections but also in flanged connections in some instances. In the latter cases, the operating conditions have mostly been severe and outside the specifications. No problems have been reported for welded connections or for seam welded pipes.

There are several possibilities to improve the crevice corrosion resistance of a piping system. One way, which is now being used in practice, is to weld overlay the flange surfaces with nickel base alloys, for example, of the Alloy 625 or C-276 types. This might be a good technical solution but long term experience is so far missing. Another possibility is to use other types of stainless steels. There seems to exist just one independent investigation in which piping systems, consisting of real components of various stainless steels, have been tested in continuously chlorinated sea water [6]. Under these realistic conditions the crevice corrosion resistance of 254 SMO flanges turned out to be at least as good as that of the most common superduplex steels, e.g. Zeron 100. For both types of steels the application limit was 0.5 ppm of residual

*654 SMO® and 254 SMO® are registered trademarks of Avesta Sheffield AB.

chlorine at a water temperature of 30°C. Furthermore, the welds were more critical for the latter steel than for 254 SMO. So changing to another type of stainless steel seems not to be an option.

However, during the last few years new, improved stainless steels have been developed and one such steel is Avesta Sheffield 654 SMO. This is a second generation superaustenitic steel, the composition of which (Table 1) is characterised by very high contents of those alloying elements most important for the resistance to localised corrosion, i.e. chromium, molybdenum and, especially, nitrogen. General descriptions of 654 SMO and its weldability are given elsewhere [7, 8]. In the present paper, the corrosion resistance of 654 SMO will be compared with that of 254 SMO and two nickel base alloys in continuously chlorinated sea water at elevated temperatures.

2. Experimental
2.1. Test Material

The test material was taken from commercially produced sheet or plate. Two stainless steels, 254 SMO and 654 SMO, were included in all tests, and the nickel base alloys 625 and C-276 in only one test. The typical chemical compositions of the alloys are shown in Table 1.

2.2. Welding

In most cases the test specimens were welded (GTAW or SMAW) using filler materials, the typical compositions of which are shown in Table 2. The welding procedures used for producing the test specimens are described in Table 3. When GTAW was used the shielding gas consisted of pure argon or, in the case of 654 SMO, of argon plus 5–10% nitrogen. The backing gas was always nitrogen plus 10% hydrogen. After welding the top sides of the welds were sand blasted and briefly treated with pickling paste. The root sides were left untreated.

2.3. Test Specimens

In test No. 1 all specimens (200 × 70 mm) were butt welded and equipped with crevice formers (ø24 mm) taken from a 2 mm thick commercial gasket material (Klinger Sil C4400) consisting of rubber bound aramid fibre. The crevice formers,

Table 1. Typical composition of tested alloys

| Alloy | Typical composition, % | | | | | | |
	C	Cr	Ni	Mo	N	Cu	Other
254 SMO	0.01	20	18	6.1	0.20	0.8	—
654 SMO	0.01	24	22	7.3	0.50	0.5	Mn 3
625	0.02	22	Bal.	9	—	—	Nb, Fe
C-276	0.01	16	Bal.	16	—	—	W, Fe

Table 2. Typical composition of filler materials

Alloy	Filler material		Typical composition, %				
	Type	Avesta	C	Cr	Ni	Mo	Other
254 SMO	GTAW	P12	0.03	21	Bal.	9.0	Nb
	SMAW	P12	0.02	21.5	Bal.	9.0	Nb
654 SMO	GTAW	P16	0.02	23	Bal.	16	—
	SMAW	P16	0.02	25	Bal.	15	—
625	GTAW	P12	0.03	21	Bal.	9.0	Nb
C-276	GTAW	—	0.02	16	Bal.	16	W

Table 3. Welding parameters

Test No.	Parent material			Welding			
	Alloy	Thickness, mm	Method	Number of Filler	Arc Energy passes	kJmm⁻¹	
1	254 SMO	3	GTAW	P12	1	1.2	
	654 SMO	3	GTAW	P16	1	1.2	
	625	1.25	GTAW	P12	1	1.2	
	C-276	3	GTAW	C-276	1	1.2	
2 and 3	254SMO	3	GTAW	P12	2	0.85–0.89	
		8	GTAW	P12-root	1	0.83	
			SMAW	P12	8	0.46–0.79	
	654SMO	3	GTAW	P16	1	1.20	
		10	GTAW	P16-root	2	0.97–1.02	
			SMAW	P16	15	0.53–0.83	

which only made contact with the parent metal, were bolted to the surface of the specimens using a torque of 40 Nm. Four parallel specimens of each alloy were exposed [9].

In test Nos 2 and 3, two kinds of specimens were used, one type being welded and exposed without crevices, the other type being unwelded but having the same type of crevice formers (18 Nm) as the specimens in test No. 1. Both types measured 100×150 mm. In each test three cold rolled (3 mm) and three hot rolled (8–10 mm) specimens were exposed [10].

2.4. Test Conditions

In test No. 1, which was performed at the CEA sea water laboratories in Normandy (France), the water was continuously chlorinated to 10 ppm of residual chlorine. The water was heated to 45°C and the test specimens were placed directly into the hot, chlorinated water. The test duration was 95 d.

Test No. 2 was performed at the sea water laboratories belonging to the Swedish Corrosion Institute. The water was continuously chlorinated to 2 ppm and heated to

55°C. As in the preceding test, the specimens were placed directly into the hot, chlorinated water and exposed for 130 d.

Test No. 3 was performed at the same laboratory as Test No. 2, and the water had the same temperature. The chlorination procedure was, however, different. Instead of placing the test specimens directly into a continuously chlorinated water, the test started with intermittent chlorination, i.e. chlorination with 2 ppm for 1 h per day. After 30 d the chlorination mode was changed to continuous chlorination with 2 ppm. The total exposure time was 100 d.

2.5. Corrosion Evaluation

After the tests the specimens were inspected in a stereo microscope at 20× magnification. If crevice corrosion was detected, the depth of attack was measured by means of a needle point micrometer or a microscope with a calibrated fine-focus knob. In the welded, crevice-free specimens, any attack (pitting) occurred in or in immediate vicinity of the weld. The depths of those pits appearing to be most severe were determined by metallographic investigations

3. Results

The results of the crevice corrosion tests are shown in Tables 4–6. The results are reported as the number of attacked crevice sites (two per specimen) and as the depth of the deepest attack. In Tables 5 and 6 no separation has been made between cold and hot rolled material. The photographs in Figures 1–4 show the deepest attack in test No. 1, and those in Figs 5 and 6 the deepest attack in Test No. 2.

Table 4. Test No. 1; creviced, welded specimens. Continuous chlorination 10 ppm, 45°C

Alloy	Crevice corrosion		Pitting corrosion	
	Attacked sites	Max. depth, mm	Attacked sites	Max. depth, mm
254 SMO	7/8	0.98	0/4	—
654 SMO	0/8	—	0/4	—
625	8/8	0.47	0/4	—
C-276	7/8	0.31	0/4	—

Table 5. Test No. 2; creviced, unwelded specimens. Continuous chlorination 2 ppm, 55°C

Alloy	Crevice corrosion	
	Attacked sites	Max. depth, mm
254 SMO	12/12	1.0
654 SMO	4/12	0.37

Table 6. *Test No. 3; creviced, unwelded specimens. Intermittent chlorination 2 ppm, 1h/24h, 55°C followed by continuous chlorination 2 ppm, 55°C*

Alloy	Crevice corrosion	
	Attacked sites	Max depth, mm
254 SMO	5/10	0.15
654 SMO	0/12	—

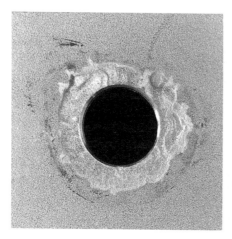

Fig. 1 *Crevice corrosion (0.98 mm) on 254 SMO in test No. 1 (1.8×).*

Fig. 2 *Crevice position on 654 SMO in test No. 1 (1.8×).*

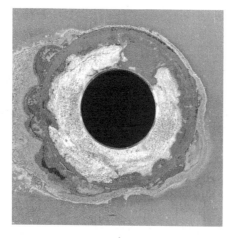

Fig. 3 *Crevice corrosion (0.47 mm) on Alloy 625 in test No. 1 (1.8×).*

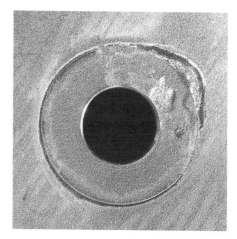

Fig. 4 *Crevice corrosion (0.31 mm) on Alloy C-276 in test No. 1 (1.8×).*

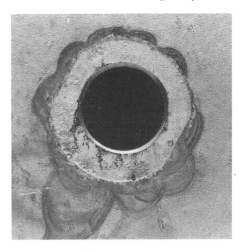

Fig. 5 *Crevice corrosion (1.0 mm) on 254 SMO in test No. 2. Hot rolled specimen (1.8×).*

Fig. 6 Crevice corrosion (0.37 mm) on 654 SMO in test No. 2. Hot rolled specimen (1.8×).

Welded specimens were included in all three tests. In test No. 1, where crevice corrosion occurred, the welds did not corrode but in test Nos 2 and 3 pitting was observed. The results of the latter tests are presented in Tables 7 and 8. The tables show the number of attacked welds (one per specimen) and the deepest pit found in the metallographic investigations. The photograph in Fig. 7 shows pitting in a 254 SMO weld and Fig. 8 shows a section through this pit. In Fig. 9, the only attack observed in welded 654 SMO specimens is shown.

Table 7. Test No. 2; welded, uncreviced specimens. Continuous chlorination 2 ppm, 55°C

Alloy	Pitting corrosion		Remark
	Attacked welds	Max. depth, mm	
254 SMO	5/6	0.5	
654 SMO	1/6	< 0.01	weld defect

Table 8. Test No. 3; welded, uncreviced specimens. Intermittent chlorination 2 ppm, 1h/24h, 55°C followed by continuous chlorination 2 ppm, 55°C

Alloy	Pitting corrosion	
	Attacked sites	Max depth, mm
254 SMO	2/5	0.6
654 SMO	0/6	—

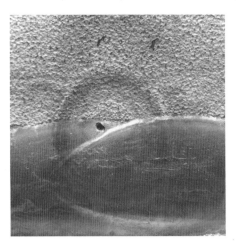

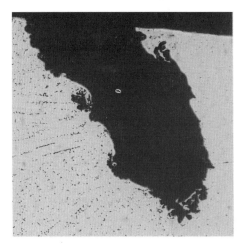

Fig. 7 Pitting corrosion in fusion line on 254 SMO in test No. 3. Cold rolled specimen (9×).

Fig. 8 Section through the pit shown in Fig. 7 (200×).

4. Discussion
4.1. Worst Case Conditions

The sea waters used in our tests have no doubt been extremely corrosive. In test No. 1 (Table 4) almost all crevice sites were attacked on the 254 SMO and the Alloy 625 and C-276 specimens and in test No. 2 (Table 5) even four out of twelve crevice sites corroded on the 654 SMO specimens. Judging from these results, the temperature limit for crevice corrosion on 654 SMO seems to exist somewhere between 45 and 55°C under worst case conditions, i.e. when the steel is immediately exposed to the hot continuously chlorinated water. It should be borne in mind, however, that chlorine levels as high as 2 ppm hardly exist in a sea water system where the water has had time enough to be heated to 55°C. For real piping systems made of 254 SMO the temperature limit for crevice corrosion in flanges is reported as 30°C at 1 ppm of chlorine [1].

In test No. 1 (Table 4) all welds have been intact. This means that crevice corrosion in the parent metal is a greater corrosion risk than pitting corrosion in the welds. In test No. 2 (Table 7), on the other hand, five out of six 254 SMO welds suffered pitting corrosion while the 654 SMO welds were resistant with the exception of a small attack in a severe weld defect. As shown in Fig. 9, the defect consists of bad penetration, leaving a 1.5 mm deep crevice between the plates. In spite of this, the attack on the weld metal in the bottom of the crevice is very small and is only revealed by a slight rust coloration. The pits in the 254 SMO welds often have a very small opening (Fig. 7) but may be quite deep (Fig. 8).

Butt welded 254 SMO tubes have been reported to be resistant to a continuously chlorinated (0.5–1 ppm) water at 40°C [11]. The fact that the 254 SMO welds were resistant at 45°C and 10 ppm (Table 4) might be due to cathodic protection being received from propagating crevice corrosion. So, because pitting did occur at 55°C the temperature limit for pitting corrosion in 254 SMO butt welds should exist somewhere between 40 and 55°C in a water containing high amounts of chlorine. Seam welded 254 SMO tubes, not containing butt welds, have been resistant at 45°C and 2 ppm continuous chlorination [12].

For 654 SMO the weld results of test No. 1 are valid because no crevice corrosion occurred at 45°C and 10 ppm of chlorine. With the exception of the weak attack in the weld defect shown in Fig. 9, all 654 SMO welds have were also resistant at 55°C and 2 ppm of chlorine. It seems, therefore, as if in the case of 654 SMO as well as in the case of 254 SMO, the weakest point in a pipe system is crevice corrosion in the flanges and not pitting in the welds. It should be stressed, however, that there must be high demands on the quality of the welds and that obvious weld defects have to be avoided.

4.2. Mild Start-up Period

The preceding discussion deals with the worst case, i.e. when the stainless steels are directly exposed to a hot, continuously chlorinated water. The conditions become quite different if, as in test No. 3, a period of intermittent chlorination precedes the continuous chloririation. As shown in Tables 6 and 8,654 SMO then becomes resistant to both crevice corrosion and to pitting corrosion in the welds. Even if 254 SMO is still attacked under these conditions, the intensity of the crevice corrosion and the number of weld attack sites are lower.

The explanation to these results might be that during the period with intermittent chlorination, the protective passive layer is growing and its resistance increasing. During the chlorination periods the environment is oxidising enough to "passivate" the steel, as in nitric acid passivation, but the duration of these periods is not long enough for crevice or pitting corrosion to initiate. An increased corrosion resistance of "aged" stainless steel has been observed by others [6, 13] who suggest that during the first weeks, a stainless steel piping system intended for chlorinated sea water, should be operated under less aggressive conditions than later on. Obviously, intermittent chlorination is a good way to start the operation.

4.3. Practical Experience

On the Oseberg A plafform, situated in the North Sea, 254 SMO is used for the sea water cooling system. After two years in service, crevice corrosion was detected in the flanges of the pipes connected to a crude oil cooler and in the threaded connections [3]. The pipe which was exposed to the hottest water, i.e. that connected to the sea water outlet, was replaced with a new pipe (12" Sch 10S × 1250 mm) made of 654 SMO. The pipe is equipped with cast flanges (welding neck RF 150#), one of which was welded to the pipe on the site by people from the plafform. The pipe was also supplied with two outlets having threaded ends.

An inspection of the pipe was made after 16 months of operation and no corrosion at all could be observed, not even in the threaded connections (Fig. 10). The pipe has been exposed to sea water containing about 0.4 ppm of chlorine. The temperature is

Fig. 9 *Section through weld with bad penetration. 654 SMO in test No. 2. Cold rolled specimen. (25×).*

Fig. 10 *654 SMO sea water pipe after 16 months of operation on the Oseberg A platform.*

normally 45°C but with regular intervals the temperature is raised to 60°C, each time for a period of 24 h.

5. Conclusions

Tests performed in hot chlorinated sea water show that:

• the crevice corrosion resistance of 654 SMO is much superior to that of the 6Mo steel 254 SMO and at least on a level with that of the nickel base alloy C-276.

• the temperature limit for crevice corrosion in 654 SMO seems to exist between 45 and 55°C in waters continuously chlorinated with high levels of chlorine.

• the temperature limit for pitting corrosion in welded 654 SMO structures seems to be higher than that for crevice corrosion but severe weld defects have to be avoided.

• if the continuous chlorination is preceded by a start-up period with intermittent chlorination, the resistance to crevice and pitting corrosion will increase considerably.

• experience from a real sea water application confirms the excellent resistance of 654 SMO in continuously chlorinated water.

6. Acknowledgements

The author gratefully acknowledges the great contribution of Dr. Damien Féron, CEA, to test No. 1 and also the financial support given by Stiftelsen Varmeteknisk Forskning to tests Nos 2 and 3.

References

1. R. Johnsen and S. Olsen, *Corrosion '92*, Paper No. 397, NACE, Houston, Tx, 1992.

2. I. Nybø, Ø. Strandmyr, Driftserfaringer fra Statfjord og Gullfaks Feltene, in Materialteknologi For Petroleumsindustrien (10–12 January,1994), NTH, Trondheim, Norway, January,1994 (in Norwegian).

3. A. Dalheim, Erfaringer med Rusffritt Stål i Process- og Sjøvannssystemer, in Materialteknologi For Petroleumsindustrien (4–6 January, 1993), NTH, Trondheim, Norway, January, 1993 (in Norwegian).

4. R. E. Lye, Materials selection philosophy for offshore production systems, in *Proc. Engineering Solutions to Industrial Corrosion Problems* (7–9 June, 1993), Paper No. 2, NACE, Houston Tx, June 1993.

5. R. Mollan, Materials engineering experiences — Snorre Project, in Materialteknologi For Petroleumsindustrien (4–6 January, 1993), NTH, Trondheim, Norway, January 1993.

6. P. Solheim, Crevice corrosion resistance of stainless steels in chlorinated sea water – laboratory testing, in *Proc. Engineering Solutions to Industrial Corrosion Problems* (7–9 June, 1993), Paper No. 10, NACE, Houston Tx, June 1993.

7. B. Wallén, M. Liljas and P. Stenvall, *Werkstoffe und Korrosion*, 1993, **44**, 83–88.

8. M. Liljas and P. Stenvall, Welding of UNS S32654 — Corrosion properties and metallurgical aspects, in *Proc. 12th Int. Corrosion Congr.* (19–24 September 1993), pp.347–359, NACE, Houston, Tx, 1993.

9. D. Féron and B. Wallén, *Corrosion '93*, Paper No. 498, NACE, Houston Tx, 1993.

10 B. Wallén, Korrosionsprovning av tre höglegerade rosffria stål i varmt klorerat havsvatten, Stiftelsen för Varmeteknisk Forskning, Materialteknik 522, Stockholm, Sweden, 1994 (in Swedish).

11. G. Nyström and S. Henrikson, Investigations of the corrosion resistance of a new high alloy duplex stainless steel in chlorinated and unchlorinated natural sea water, in *Proc. 11th Scand. Corrosion Congr.* (19–21 June, 1989), Høgskolesentret i Rogaland, Stavanger, Norway, June 1989.

12. B. Wallén and S. Henrikson, *Werkstoffe und Korrosion*, 1989, **40**, 602–615.

13. P. O. Gartland and J. M. Drugli, *Corrosion '91*, Paper No. 510, NACE, Houston, Tx, 1991.

14

A New Generation of High Alloyed Austenitic Stainless Steel for Sea Water Systems: UR B66

J.-P. AUDOUARD and J.-C. GAGNEPAIN

Creusot-Loire Industrie, Research Centre for Materials, BP 56, 71202 Le Creusot, France

ABSTRACT

A new 6Mo super-austenitic stainless steel has been developed as a result of investigations carried out on a series of high nitrogen alloys containing high molybdenum, chromium, nickel and complementary additions of tungsten and copper. These investigations showed that the highest alloyed Mo–N grades achieve a very high corrosion resistance, including pitting and crevice, but that their structural stability appeared to be insufficient to permit the development of industrial products. Tungsten addition has proved very useful from a corrosion point of view in replacing a certain amount of Mo while the structural stability was found to be better than for the highest Mo grades.

This paper discusses the properties of the new high nitrogen-containing UR B66 stainless steel with special emphasis on structural stability, mechanical properties and corrosion performance in chloride-containing media. From pitting and crevice corrosion tests on unwelded and welded specimens carried out in oxidising chloride-containing media and in sea water, the new UR B66 grade appears to be one of the best candidates for use in natural and treated sea water.

1. Introduction

In general, only type 304 and the lower alloyed stainless steels (SS) are prone to localised corrosion in artificial neutral chloride-containing solutions at ambient temperature. The situation is quite different in natural sea water where the open circuit potential of stainless steels is displaced from +300 to +450 mV/Ag–AgCl depending upon the climatic conditions. As most of the usual stainless steels including 316 grades have pitting potentials in the region of +100 to +300 mV/AgCl, they are prone to pitting attack, or to crevice corrosion. Hernandez *et al.* [1] showed that these high values of the electrochemical potential were related to the formation of a biofilm which appears in slowly moving water, and which can, very markedly, depolarise the cathodic process. Biochemical activity, involving several species of micro-organisms, seems to be the driving force for the corrosion processes. As biological activity is much dependant upon local climatic conditions, it can be inferred that corrosive conditions could vary in a significant manner depending upon the geographical location.

Audouard *et al.* [2] demonstrated that medium grade austenitic SS containing 4% molybdenum, ≥ 20 % chromium and nitrogen or 25Cr duplex alloys are much more resistant than 316 and 317 grades, so that they are generally able to withstand localised corrosion in natural sea water provided that these steels do not contain manganese sulfides. This can be achieved if the sulfur content is lower than about 0.002 % in the bulk metal.

Unfortunately, macro-organisms develop inside sea water circuits and use the biofilm as a nutrient. To avoid this detrimental process, which tends to choke pipes, oxidising chlorinated products are added to sea water. If overdosed, these products drive the open circuit potential of SS to high values that favour pit initiation and crevice propagation. The most common way to fight localised corrosion consists in improving the pitting and crevice resistance of stainless steels by adding sufficient amounts of alloying elements like chromium, molybdenum and nitrogen which give a better resistance mainly to pit initiation [3] while copper and nickel are recognised for improving pit propagation and crevice corrosion.

The 6% Mo austenitic grades can be used in such conditions, but high chromium contents (22–25%) and nitrogen additions are required in the most aggressive environments. Unfortunately, the structural stability of such materials is generally insufficient to permit the development of thick industrial products.

The new superaustenitic UR B66 alloy, which has been developed to provide a high corrosion resistance close to that of nickel base alloys, high mechanical properties and a good structural stability, appears to be a good candidate to solve the problem.

2. The UR B66 Alloy — Chemical Composition and Structural Stability

During the last fifteen years, stainless steel grades with high chromium (21–25%) and molybdenum (5– 6.5%) contents have been introduced. Such development has been possible as a result of new technical capabilities in increasing (0.15–0.25%) and controlling the nitrogen level of steels. Indeed, this alloying element has appeared to be one of the most important factors in this development because

(1) it exerts a powerful effect on the localised corrosion resistance,
(2) it improves the mechanical properties, and
(3) it stabilises the austenitic structure thus preventing intermetallic phase precipitations.

From a metallurgical point of view, the improvement of the austenite stability has allowed an increase in alloying element contents that are beneficial to the corrosion resistance while maintaining the other properties.

Taking into account these results, experimental high nitrogen super-austenitic heats containing high molybdenum (4.5–7%), chromium (21–25%), nickel (17– 25%) and complementary additions of tungsten (0–2%), copper (0–1.5%) and manganese (1–6%) were produced. Laboratory investigations carried out on these materials [6] showed that a very high corrosion resistance including pitting and crevice resistance can be achieved (Figs 1 and 2).

Nevertheless, the structural stability of alloys containing 7% Mo and 0.5% N appeared to be insufficient to develop industrially thick products, but tungsten

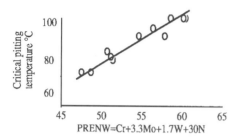

Fig. 1 Chemical composition and critical pitting temperature correlation (PRENW)[6]. ASTM G 48A– 6% FeCl₃.

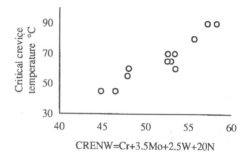

Fig. 2 Chemical composition and critical crevice temperature correlation (CRENW)[6]. ASTM G78A–6%FeCl₃.

additions have been found to be beneficial for the corrosion resistance and not detrimental for structural stability [4, 5] has shown in Fig. 3.

As a result of these conclusions, a composition for the new 6 Mo super-austenitic grade UR B66 was developed. Table 1 shows the typical analysis compared with other austenitic and Ni-base alloys commonly used in oxidising chloride solutions. The Pitting Resistance Equivalent values including the positive effect of W on localised corrosion resistance (PRENW) [6] are also given.

The alloying elements were carefully balanced in order to promote an austenitic primary solidification for preventing the formation of ferrite islands which are the areas most prone to sigma phase precipitation during cooling. The combination of a high nickel (22%) with 25% chromium, tungsten addition and high nitrogen contents (0.45% by weight) lead to a fully austenitic microstructure after quenching from 1150°C with no indication of intermetallic (sigma, chi, ..) carbide or nitride precipitations.

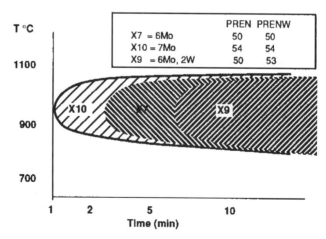

Fig. 3 Effects of molybdenum and tungsten additions on the temperature time precipitation diagrams of Fe22Ni–24Cr–3Mn–0.45N SS [4, 5].

Table 1. Chemical compositions of grades tested (wt %)

Designation	UNS	Structure	C	Cr	Ni	Mo	Cu	N	Other	PRENW
316L	31 603	austenitic	0.03	17	12	2.2				25
317LMN	S31726	austenitic	0.02	18	14	4.5		0.14		35
UR B26	N08926	superaustenitic	0.02	20	25	6.3	1	0.2		44
UR B66	—	**superaustenitic**	**0.02**	**24.5**	**22**	**5.6**	**1.5**	**0.45**	**W = 2**	**54**
UR 625	N 06625	nickel base	0.02	21.5	60	9			Nb = 4	51
Alloy C276	N 10276	nickel base	0.01	16	Bal.	16				69
Alloy C22	N 06022	nickel base	0.01	21	Bal.	13			W = 3	64
UR 45N	31 803	duplex	0.02	22	5.7	2.8		0.12		35
UR 52N+	32 550	duplex	0.02	25	6.5	3.5	1.5	0.24		40

3. Mechanical Properties

The mechanical properties of the UR B66 grade are presented in Table 2. The alloy exhibits high mechanical properties at ambient as well as at low temperature.

The Yield Strengh value (YS 0.2%) is > 450 MPa and the Ultimate Tensile Strengh value (UTS) is > 800 MPa for hot rolled plates and heavy bars. The toughness is very high (KV > 300 J) for hot rolled plates and bars Ø 235 mm. Thicker bars still exhibit KV toughness values > 100 J even for samples taken from the center of a 700 mm Ø bar. At –50 and –196°C, the mechanical properties of the alloy UR B66 remain very high for both tensile test results and Charpy V test results.

Those very high YS and UTS combined with excellent ductility and toughness, make the alloy very attractive for design purposes.

Table 2. Typical mechanical properties obtained for UR B66

Products type	Size	Testing T°C	Tensile properties			Toughness
			Y.S (MPa)	U.T.S. (MPa)	E%	Charpy V (J)
Hot roll plates	12 mm	20	446	850	60	> 300
		–50	—	—	—	> 300
		–196	—	—	—	< 250
Forged products (bars)	Ø 235 mm	300	476	863	80	
		20	—	—	—	> 300*
		–50	585	1003	80	> 300*
		–196	—	—	—	> 250*
	Ø 500 mm	20	—	—	—	271*
		–50	—	—	—	239*
	Ø 700 mm	20	—	—	—	151*
		–50	—	—	—	127*

* samples issued from the centre of the product.

4. Weldability

UR B66 can be welded using GTAW, GMAW or SMAW processes with covered electrodes. Typical electrodes or filler metal to be used for sea water applications, including treated sea water, are nickel based products such as alloy C22, alloy 59 or equivalent. Homogeneous filler metal can be used for specific applications instead of Ni base materials; this can be the case in oxidising neutral chloride-containing solutions, where Ni base alloys suffer from transpassive dissolution.

The weldability appears to be better than that of the 6 Mo super-austenitic SS. The alloy is even less sensitive to hot cracking or phase precipitation effect. Welding without filler metal can also be utilised without excessive loss of corrosion resistance properties. This is possible due to the high structure stability of the grade.

Some corrosion test results carried out on welded samples are presented below.

5. Corrosion Performances of UR B66 in Synthetic Chloride-Containing Media
5.1. Testing Procedures

AISI 316L and higher alloyed SS are recognised to be perfectly resistant to pit initiation during immersion tests conducted at ambient temperature in aerated 30 gL^{-1} NaCl neutral solution which represents the average chloride concentration of sea water, provided their sulfur content is sufficiently low (generally < 0.003%). Nevertheless, pitting can occur when temperature increases. The effect of temperature on the resistance to pit initiation was evaluated by measuring the pitting potential, E_p, taken from the polarisation curves.

In some circumstances, the geometry of pipes or vessels can lead to local stagnant conditions which promote chloride enrichment and acidification [2]. In this situation, the localised corrosion can propagate either by local depassivation of the crevice wall, i.e. by new pits occurring inside the crevice or by uniform dissolution of the entire internal surface of the crevice.

The resistance to the initiation of new pits inside the crevice can be evaluated by measuring the pitting potential E_p taken from polarisation curve plots. The resistance to propagation by uniform dissolution can be investigated by measuring the maximum current density, I_{max}, of the active peak of the same curve. The tests presented in this paper were conducted in a 300 gL^{-1} NaCl solution acidified to pH 1 which is fairly representative of the medium inside a crevice. The electrolyte was heated to 60°C in order to increase the aggressiveness.

The conditions encountered in sea water treated by oxidising biocides are generally simulated using the well-known FeCl$_3$ solution (ASTM G-48). The materials were ranked by measuring their Critical Pitting Temperature (CPT) according to ASTM G48 and Critical Crevice Temperature (CCT) according to ASTM G78 on welded and non-welded specimens. In the case of welded samples, the washers were fitted near the weld to promote the crevice in the heat affected zone.

Finally, the effect of chlorination on the biofilm development and the consequences on the corrosion processes were investigated on 316L and 25Cr alloys (austenitic and duplex respectively). The tests were carried out in a northern marine station (Brest, France) in the following conditions:

(1) on polished coupons in a loop fed by renewed natural sea water with a controlled circulation rate of 1 ms^{-1}, and
(2) on samples fitted with multi-crevice washers, as described previously, immersed in large tanks fed by renewed natural sea water.

The rest potential evolution of the two types of samples was also recorded

5.2. Materials Tested

The typical compositions of the tested materials are presented in Table 1. The coupons used for the tests were cut from plates or from welded joints. Unwelded materials were prepared by polishing with 180 grit paper. Welded samples surfaces were prepared by pickling.

Unwelded specimens are indicated with their UNS number if this exists, and welded joints with UNS number followed by W. The welded joints tested are presented in Table 3.

The welded joints tested were prepared using the shielded metal arc (SMAW) process. This process is the most common and appears very conservative in evaluating the mechanical and corrosion properties of welded structures.

The welds were made on 5–9 mm thick plates. The welding was done in several passes with overalloyed filler materials (Table 3). UR B66 has been welded with C22 type filler material. X-ray examinations of the welded joints showed the welds to be in agreement with ASME specifications.

5.3. Pitting Corrosion

5.3.1. Electrochemical tests in NaCl 30 gL^{-1}
These tests were conducted on unwelded specimens. The behaviour of the different

Table 3. Welded joints and parent metals tested

Specimen name	Parent material	Filler material	Welding process
S31726 W	S31726	Alloy 625	SMAW
N08926 W	N08926	Alloy 625	SMAW
B66 W1	B66	Alloy 22	SMAW
N06022 W	N06022	Alloy 22	SMAW
N06625 W	N06625	Alloy 625	SMAW .

Type	Cr	Ni	Mo	Cu	N	Other
Alloy 625	21	Bal.	9	—	0,14	
Alloy 22	22	Bal.	13			W = 3

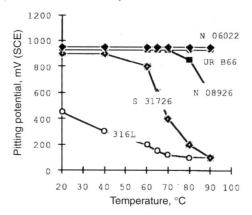

***Fig. 4** Effect of temperature on the pitting potential in NaCl 30 gL⁻¹ – ASTM G61.*

materials is clearly shown to be dependant on the composition of the SS, particularly at high temperature (Fig. 4) since the highest alloyed materials exhibit the highest pitting potential values. Among the best materials, one must stress that the UR B66 super-austenitic grade appears to be as resistant to pit initiation as the N 06022 Ni base alloy up to at least 95°C .

5.3.2. *Critical pitting temperature in FeCl₃*

The critical pitting temperatures (CPT) have been determined by optical observations and weight loss measurements with a weight loss criterion of 10 mg on standard samples. The CPT are summarised in the graph presented in Fig. 5.

The results obtained on the unwelded materials improve, as expected, with their PREN level from 35°C (95°F) for the S31726 to 70°C(158°F) for the N08926 alloy. The UR B66 grade exhibits a CPT higher than 100°C (212°F), superior to that of the nickel alloy N06625.

The critical pitting temperatures of the welded samples are found to be slightly lower than for unwelded samples, but the general ranking of the materials remain about the same as for unwelded materials.

An important point is that the UR B66 grade exhibits a CPT value much higher than the conventional 6Mo superaustenitic (UNS08926) and even the Ni-base alloy N06625 in welded conditions. This can be attributed to an improved alloying balance between Cr and Mo and to the beneficial effect of W and N additions which lead to a superior corrosion resistance and to an improved structural stability.

5.4. Crevice Corrosion and Pit Propagation

5.4.1. *Electrochemical tests*

Potentiodynamic curves have been plotted in an acidic concentrated medium simulating the crevice media (NaCl 300 gL⁻¹, pH = 1, T = 80°C). These curves are presented in Fig. 6. The maximum dissolution current taken from these curves gives an estimation of the maximum propagation rate inside the crevice, while the rupture

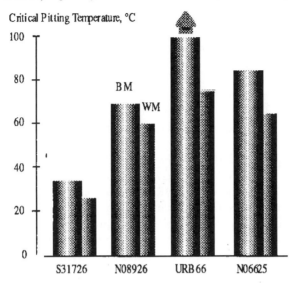

Fig. 5 *Critical pitting temperatures measured on welded and welded specimen. ASTM G-48 Test. BM = Base Material, WM = Welded Material.*

potential gives an indication of the ability of the crevice area to propagate by formation of new pits.

In these conditions, the 6 Mo grade (UNS 08926) exhibits an important activation peak and a rupture potential around 200 mV (SCE). The UR B66 grade and the nickel base alloy N10276 present roughly the same behaviour, with a low activation peak and a rupture potential around +800 mV (SCE). These last results demonstrate the outstanding crevice corrosion resistance of the new UR B66 alloy.

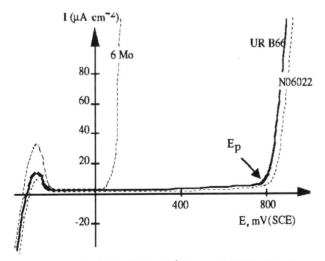

Fig. 6. *Potentiokinetic curves in NaCl 300 gL⁻¹ deaerated, 80°C, pH = 1.*

5.4.2. Critical Crevice Temperature (CCT)
To simulate the crevice phenomenon, PTFE multi-crevice washers were fitted on the samples with a controlled torque of 0.28 Nm. In the particular case of welded samples, the washers were fitted near the weld in order to promote the crevice in the heat affected zone.

The crevice corrosion sensitivity was investigated by measuring the critical crevice temperatures (CCT) supplemented by optical observations and by weight loss measurements. The results of the tests are presented in Fig. 7 for unwelded and welded samples.

In unwelded conditions, the CCT of the grade S31726 is around 25°C. Considering the super-austenitic and nickel base alloy families, the N08926 specimen exhibits a CCT around 50°C (122°F) whereas the URB66 presents a CCT of about 70°C (158°F). In the testing conditions the N06625 specimen exhibits a CCT lower than 45°C (113°F).

The CCT values of the welded samples are found to be rather lower than those measured on unwelded sheets but the ranking of the alloys remain the same.

These results show that the crevice corrosion behaviour of the UR B66 grade outperforms the 6 Mo grade and even N06625. Only the Ni base alloy N06622 is found to be superior.

6. Corrosion Performances of Stainless Steels in Sea Water
6.1. Behaviour of Stainless Steels in Natural Sea Water

In general, the shape of the rest potential curves vs time of SS immersed in natural

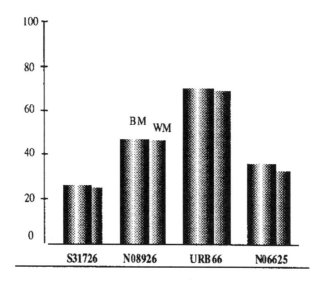

Fig. 7 Critical crevice temperatures measured on unwelded and welded specimen. ASTM G-78 Test.

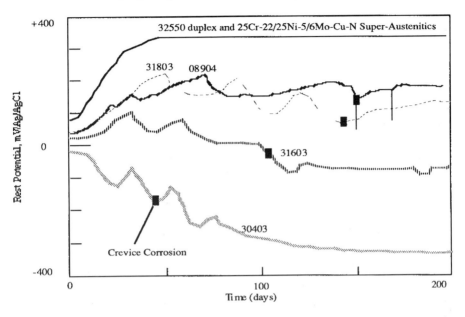

Fig. 8 *Long term immersion tests in natural sea water. Rest potential evolution vs time for various SS [2].*

sea water with a low velocity (1 ms^{-1}) can be ranked in four classes depending on the corrosion resistance of the steels tested [2] as shown in Fig. 8.

1. Steels whose potentials fall very rapidly from the outset, as exemplified by UNS 30403 grade (AISI 304L). Pits and crevices could be observed to form rapidly enough after a few weeks and to become very pronounced after about three months, to such an extent that the samples are usually perforated.

2. UNS 31603 grade (AISI 316L) is typical of the second group whose potentials fall more slowly. The time before corrosion perforates the samples is increased to one or several months.

3. The third class comprises alloys like the duplex UNS 31803 and the austenitic UNS 08904 which display at first a marked increase of potential. This initial high potential period is followed by some intermittent potential drops corresponding to the onset of corrosion followed by some repassivation. The incubation time is thus markedly longer than it is for the previous materials. The crevice corrosion appeared after several months or 1 year. This supports the hypothesis according to which a high chromium content improves resistance to pit initiation while resistance to propagation must be obtained by additions of nickel, molybdenum and copper.

4. The fourth and final group is made up of super-duplex and super-austenitic alloys with high Cr, Mo and N contents (UNS 32550 and 25Cr–22/25Ni–5/6Mo–1.5Cu–N super-austenitics). The rest potential measured on these steels

is very high and remains constant at high values near +300 mV/Ag–AgCl). These materials were found to be extremely resistant in natural sea water after more than one year exposure without any visible attack and any potential fall. Such a behaviour can be attributed to a very good resistance both to pit initiation and to propagation.

The general rest potential evolution and the corrosion behaviour of SS seem to remain valid whatever the climatic conditions of exposure—as demonstrated in a Paneuropean Research published elsewhere [7]. This means that the depolarisation process due to the biofilm adhesion and growth is present whatever the geographical location. Nevertheless, the initial rest potentials were found to be about 150 mV lower in the northern regions (Norway Sea and 48°N Atlantic) than in the southern station (Mediterranean) and the potential ennoblement rate higher as the initial potential was lower. Such a behaviour is valid whatever an SS is considered, i.e. 316L, medium range duplex (UNS 31 803) and 25Cr super-duplex and super-austenitic SS.

No clear correlation was found in this research between the initial rest potential or the rest potential ennoblement evolution and the crevice corrosion behaviour, apart for the 316L SS, which was much more corroded in the southern station. The interpretation of these phenomena is not easy but, from a practical point of view, it is very important to stress that even if climatic conditions exert some effect on the agressiveness of sea water, SS with a sufficient PREN or PREWN value (equal or > 40) were never found to be corroded during these tests.

In such conditions, it is obvious that the new UR B66 alloy, with a PRENW value > 50 is to be considered as a very valuable material for application in natural sea waters.

6.2. Effect of Chlorination on Crevice Corrosion in Sea Water

The rest potential values recorded on polished coupons and on multi-crevice corrosion coupons were found to be similar. The curves plotted on UNS 31603 and 25Cr–22/25Ni–6Mo–Cu–N Super-Austenitic alloys are shown in Figs 9 and 10 respectively.

The increase in the initial value of the rest potential depends on the residual chlorine concentration. This means that at least during the first period of immersion the SS act as a redox indicator. After that, the rest potential evolution depends on the crevice corrosion of the materials.

Thus, in the case of the 31603 SS, the rest potential increased very slowly in natural sea water and reached +250 mV and then crevice corrosion occurred (65% of the potential crevice spots were corroded). With 0.5 ppm residual chlorine, high potential values were reached quickly followed by potential drops due to crevice corrosion with 100% of the crevice spots corroded. With 0.2 ppm residual chlorine, the highest rest potential value was about +50mV followed by a plateau around –100 mV; in these conditions, crevice corrosion was also observed but only 20% of the potential crevice spots were corroded.

In the case of the 25Cr–22/25Ni–6Mo–Cu–N super-austenitics which have been

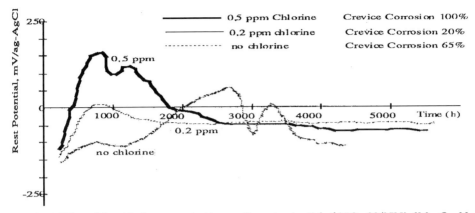

Fig. 9 *Effect of the chlorine concentration on the rest potential of 25Cr–22/25Ni–6Mo–Cu–N super-austenitic alloys.*

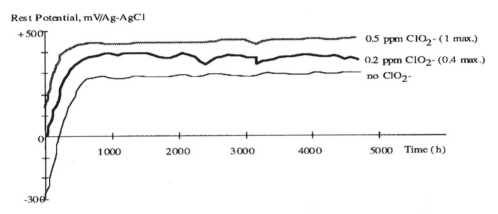

Fig. 10 *Effect of the chlorine concentration on the rest potential of 25Cr–2V25Ni–6Mo–Cu–N super-austenitic alloys.*

found to be perfectly resistant either in natural sea water or in sea water chlorinated up to 0.5 ppm residual, the potential plateau is situated at around +300 mV without chlorination and around +350 mV with 0.2 ppm chlorine and +450 mV with 0.5 ppm chlorine.

7. Conclusions

The results of the corrosion tests presented above show that the 25Cr–22/25Ni–6Mo–Cu0.2/0.45N super-austenitic alloys are very resistant to localised corrosion in chloride-containing media and in sea water. Such materials withstand crevice corrosion in treated sea water up to at least 0.5 ppm residual chlorine at ambient temperature. In the most agressive conditions, such as hot, acidic, oxidising, chloride-containing solutions, the combination of an accurate balance between Cr and Mo, and N and W additions give the new UR B66 grade an outstanding corrosion resistance and an improved structural stability — factors which account for the good resistance of welds. The localised corrosion performance of this grade appeared to be better than that of the UNS 06 625 alloy and close to that of the UNS 06022 alloy.

References

1. G. Hernandez, C. Lemaître, G. Beranger and J.-P. Audouard, Biological microfouling formation and corrosion of 316L Stainless Steel: role of poison alloying elements, *Int. Symp. on Corrosion and Corrosion Control for the Off-Shore and Marine Construction*, Xiamen-Amoy, China, 6–9 September 1988.

2. J.-P. Audouard, A. Desestret, D. Catelin and P. Soulignac, Special stainless steels for use in sea water, *Corrosion '88*, Paper No. 413, NACE, Houston, Tx, 1988.

3. R. E. Malpas, P. Gallagher and E. B. Shone, *in Proc. Chlorination of Sea water Systems and its Effect on Corrosion*, 4 March, 1986, Paper TRCP.2919, Roy. Soc. Chem Ind., 1986, Birmingham UK.

4. J. Charles, F. Dupoiron, J.-C. Gagnepain, R. Cozar, B. Mayonobe, A new High nitrogenaustenitic stainless steel with improved structure and corrosion resistance properties, (NITO 1994, Oslo, Norway).

5. J-C. Gagnepain, F. Dupoiron, M. Chavet, R. Cozar and B. Mayonobe, Aciers inoxydables austénitiques à hautes caractéristiques mécaniques et haute tenue á la corrosion (Cercle d'étude des métaux, Mai 1994, Saint-Etienne, France).

6. F. Dupoiron, M. Verneau, J. Charles, R. Cozar, and B. Mayonobe, Nouveaux aciers inoxydables superaustenitiques à fortes teneurs en azote et hautes performances (2ème Colloque Européen "Corrosion dans les usines chimiques et parachimiques", Grenoble, 21/22 Septembre 1994, France).

7. J.-P. Audouard, C. Compère, N. J. E. Dowling, D. Féron, D. Festy, A. Mollica, T. Rogne, V. Scotto, U. Steinsmo, K. Taxen and D. Thierry, Effect of marine biofilms on stainless steels. Results from a European Exposure Program. *Int. Conf. on Microbially Influenced Corrosion*, 8–10 May, 1995, New Orleans, La, USA.

15

Cathodic Protection of Stainless Steels in Chlorinated Sea Water and other Saline Water Solutions — A New Approach by Use of the RCP Method

J. M. DRUGLI, T. ROGNE, P. O. GARTLAND* and R. JOHNSEN*

SINTEF Corrosion and Surface Technology, N-7034 Trondheim, Norway
*CorrOcean as, Teglegården, N-7005 Trondheim, Norway

ABSTRACT

Corrosion failures have been experienced in sea water systems made of highly alloyed stainless steels in the offshore industry. This paper presents the protection method called Resistor-controlled Cathodic Protection (RCP) as a means to avoid the problem in a simple and economic way. The principles of the method are outlined and its use in various environments is discussed. Cathodic polarisation data obtained by long term testing of a highly alloyed stainless steel in chlorinated sea water are presented. Reliable data of this type are needed for the design of RCP systems.

The RCP method is now established as a proven technology and the system is marketed worldwide by CorrOcean a.s. and SINTEF Corrosion and Surface technology.

1. Background

The application of corrosion resistant materials, including highly alloyed stainless steels for sea water systems, has increased rapidly in the last decade. No materials, not even stainless steels, are resistant under all environmental conditions. In design, there is a tendency to venture as close as possible to the environmental limits of the material, because the alternative usually is to choose a more expensive material. When a material with corrosion resistance limits close to the service conditions is selected, there is always a risk that the corrosion limitations of the material can be exceeded due to service and environmental conditions which are more severe than foreseen, or the material specified has unexpected weaknesses or a larger scatter in quality than anticipated.

An example of this type of material selection is the application of austenitic 6% molybdenum steels (6 Mo) in chlorinated sea water systems in the offshore industry. Based on laboratory tests, mainly covering pitting corrosion, these highly alloyed stainless steel qualities were found to be corrosion resistant in chlorinated sea water up to temperatures of 30–35°C and 1.0–1.5 ppm free residual chlorine. In practical applications, however, crevice corrosion on this type of material in sea water systems has been experienced at temperatures far below the design temperatures applied [1, 2].

The temperature/potential limits determined in research work at SINTEF for a good quality material of 6 Mo stainless steel is given in Fig. 1 [3]. In chlorinated sea water the potential is in the region of 500–600 mV (SCE). As seen from the Figure a service temperature of 30–35°C should be safe with respect to pitting corrosion on base material or welds. However, the minimum critical crevice temperature (CCT), which is difficult to determine exactly due to the large number of variables which are influencing the result, is found to be below the design temperatures applied, i.e. below 30°C, even for materials with satisfactory microstructure.

The corrosion problems experienced in stainless steel sea water systems of this type require preventive methods to stop the corrosion. An alternative to changing to a more corrosion-resistant material, which is very expensive (up to several million NOK for an offshore sea water system), is to apply a protection method called RCP (Resistor-controlled Cathodic Protection), which is a simple and cost-effective way to handle the problem. In the present paper the principle of the method, possible areas of application, and some new results important for design of RCP systems will be presented.

The RCP method is based on a patent claim from SINTEF: entitled "Method and arrangement to hinder local corrosion and galvanic corrosion in connection with stainless steels and other passive materials". The method is now being commercialised

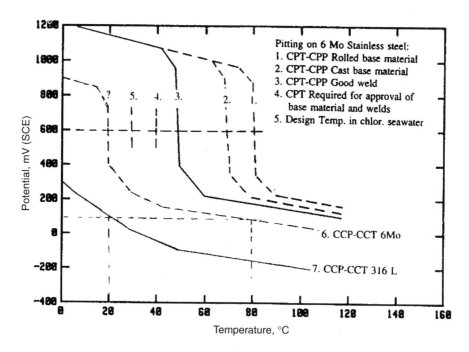

Fig. 1 *Critical pitting potentials (CPP) for a 6 Mo stainless steel vs temperature determined by long duration testing in 3% NaCl, sea water, and 6 % FeCl$_3$ solution. The critical crevice potentials (CCP) for both 6 Mo and AISI 316 L are also indicated. CPT = Critical Pitting Temperature. CCT = Critical Crevice Temperature.*

in cooperation between SINTEF Corrosion and Surface Technology and CorrOcean as. The first RCP systems are already in service [1].

2. Critical Potentials for Corrosion Initiation and Repassivation

In chlorinated sea water the potential of passive stainless steels rises to potentials about 600 mV (SCE) within a few hours. At these high potential levels local corrosion initiation is more likely than at lower potentials. The dependence of critical pitting and crevice corrosion potentials for corrosion initiation on temperature in sea water for a 6 Mo stainless steel are indicated in Fig. 1. Crevice corrosion is often the most critical form of corrosion for this type of material. Figure 1 also indicates the critical crevice potentials (CCP) temperature dependence for a 6 Mo and a 316 type stainless steel. The Figure indicates an upper safe temperature of about 20°C in chlorinated sea water for 6 Mo materials (at 600 mV (SCE)). If the material is of poor quality and/or the crevices are very narrow, as in threaded connections, the safe temperature may be even lower. Type 316 stainless steel will not be safe in chlorinated sea water even at 0°C.

The figure indicates that if the potential is lowered to about 100 mV (SCE) the safe upper temperature increases to about 80°C for 6 Mo SS and to about 20°C for 316 SS.

However, even for a smaller reduction of potential the probability for corrosion initiation is considerably reduced at the temperatures indicated.

For ongoing corrosion in crevices a lowering of the potential to 100 mV (SCE) will reduce the corrosion rate, but for repassivation to occur the potential has to be lowered somewhat further. Repassivation potentials for 6 Mo SS have been measured to be – 125 mV (SCE) at 60°C [4] and to about –200 mV at 80°C [5]. New results, obtained at SINTEF, indicate that for deep crevices (about 10 mm) in 2" flanges the repasssivation potential may be somewhat lower especially for relatively short times at the potential (≤ 24 h) [6].

3. Resistor-controlled Cathodic Protection (RCP)

3.1. The RCP Principle

The RCP method is developed to prevent local corrosion on stainless steels in piping systems with various types of saline waters, in which a critical combination of potential and temperature may be exceeded. Galvanic corrosion on materials connected to stainless steel can also be hindered. The method has been described previously elsewhere [7, 8], and also in the present volume in Chapters 5 and 11.

The basic principle of the method is to apply a cathodic current to a pipe system of stainless steel using a resistor in series with an anode to control both the potential on the stainless steel and the anode current output. The principle is shown schematically in Fig. 2. The voltage drop over the resistor is designed to obtain sufficiently, but not excessively negative, polarisation of the steel. In this way the potential of the steel is kept in the potential protective range, where the current requirements are very small in many environments, e.g. in chlorinated sea water, natural sea water above 30–40°C, produced water in connection with oil and gas processing and many other saline solutions.

Due to very low current requirements in the relevant potential range, each anode can protect large lengths of a pipe system with a very low anode consumption rate. The current requirements in chlorinated sea water at temperature 30°C for certain test conditions are shown in Fig. 3. The current requirements for chlorinated sea water and other types of saline waters are further discussed later in this paper.

3.2. Resistor Control of Anode Current Output

A sacrificial anode of a zinc or an aluminium alloy directly coupled to the pipe system will polarise the nearby stainless steel to much more negative potentials than required. This will also increase the current requirements considerably and lead to a very rapid anode consumption. The clue is therefore to control the potential by use of a resistor, or a resistor in series with a diode, as indicated schematically in Fig. 2. With R as the total resistance in series and I the current output from the anode, the potential on the stainless steel pipe near the anode, E_n, becomes:

$$E_n = E_a + IR$$

where E_a is the anode potential.

For the more distant parts of the pipe the potential, E_f, becomes:

$$E_f = E_n + \Delta E_{sea\ water}$$

where $\Delta E_{sea\ water}$ is the potential drop in the sea water between near and far positions.

By suitable choice of the total resistor value, which control I, the CP system can be designed such that both E_f and E_n are only slightly negative to the safe potential for the material, where the current load is low.

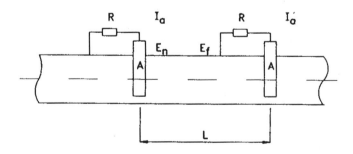

Fig. 2 *Schematic of the RCP method applied to a piping system.*

3.3. Potential Variations Along the Pipe

In many environments, e.g. in chlorinated sea water, the required current density to reduce the potential below the safe limit for corrosion initiation, is very low. High conductivity combined with low current densities makes it possible to protect relatively long pipe sections with a small number of anodes. The maximum spacing between the anodes will be determined from calculations of the potential profile along the pipe utilising boundary conditions like those given in Fig. 3.

An example of such calculations is shown in Fig. 4. This calculation was made with the computer program GALVCORR [7] In the present calculation example for a 300 mm pipe and a chlorination level of 0.5 ppm, the distance between the anodes was somewhat arbitrarily set to 50 m, while the value of the resistor was varied. The 150 W resistor gives a value of E_f slightly more positive than +100 mV (SCE).

As an example it can be mentioned that typical spacings between anodes for 100 and 250 mm 6 Mo SS pipes in chlorinated sea water with about 0.5 ppm free residual chlorine and temperature 30°C, are calculated to be 28 and 44 m, respectively. The spacing between the anodes, however, will vary dependingon the type of electrolyte in the system, on temperature, the type of stainless steel, the potential required to secure protection, eventual branching of the pipes, closed or open valves and current drain to vessels and heat exchangers, etc.

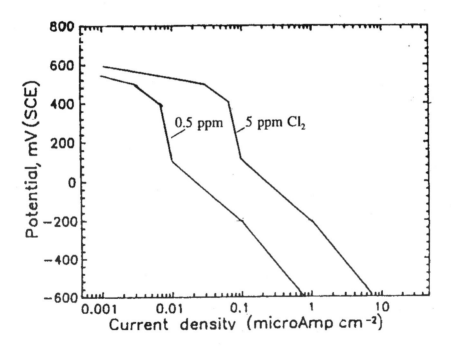

Fig. 3 *Smoothed cathodic current density curves from potentiostatic tests for stainless steels in chlorinated sea water in dependence of the potential at 30°C, and the chlorination level 0.5 ppm. The specimens were pre-exposed at zero current potential for 2–5 d before start of polarisation.*

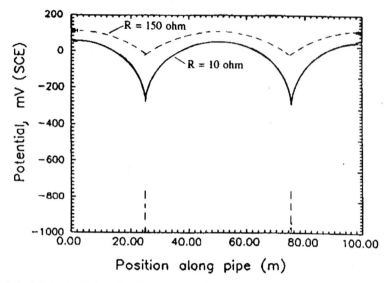

Fig. **4** *Calculated potential profiles for a 300 mm diameter pipe of 6 Mo stainless steel carrying chlorinated sea water at 30°C. RCP anodes at the position 25 and 75 m. Resistor values are 150 Ω (– – – –) and 10 Ω (———).*

3.4. Anode Consumption Rate

Based on calculations similar to those shown in Fig. 4, it is possible to find the maximum spacings between the anodes and also suitable values for resistors that result in sufficient protection and long lifetimes of the anodes. The anode currents I_a with 10 and 150 Ω in series with the anodes will be 70 and 6.7 mA, and the corresponding anode consumption for a lifetime of 10 years will be 11 and 1.0 kg, respectively.

Even for moderate anode weights the life time with suitable resistor values can be several years for a practical system. For the example above with anode spacings of 28 and 44 m for 100 and 250 mm pipes and for maximum potential $E_f = +100$ mV, the anode consumption for a lifetime of 10 years is calculated to be 0.4 and 1.6 kg, respectively.

However, the anode consumption rate will vary depending on the type of electrolyte in the system, on temperature and also on the type of stainless steel and the potential required to secure protection.

3.5. Anode Design

Several type of anodes can be applied. Because of the low anode current output required, the anodes may be mounted in special fittings and the anodes can be located outside the pipes. This types of anodes will not disturb the pipe flow, but the system has to be closed down for replacement of the anode. However, anodes for chlorinated sea water systems have been designed for long lifetimes varying from 5 to 30 y. Figure 5 gives an example of an anode mounted in a special fitting for protection of a valve and the connected pipe.

Anodes can also been mounted in high pressure access fittings as shown in Fig. 6. The access fitting and retriever is standard product of CorrOcean a.s. The anode lifetime can be 1–5 y and the anode can be replaced under normal operation of the pipe system.

The anodes are electrically insulated from the pipes by an insulating material which also is designed to prohibit fragments of the anodes falling off and mixing with the main stream.

4. Cathodic Properties of Stainless Steels in Chlorinated Sea Water and Other Saline Waters

For design of a RCP system the required cathodic current density for polarisation to a certain safe potential level has to be known. The data available with respect to chlorinated sea water are those obtained by laboratory testing performed at SINTEF Corrosion and Surface Technology. The results are based on two potentiostatic tests,

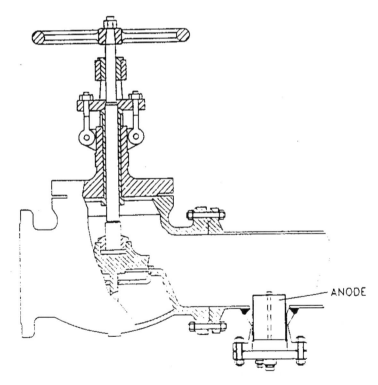

Fig. 5 *Schematic drawing of a long life anode mounted in a special fitting on a pipe close to a valve.*

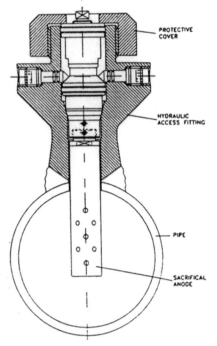

Fig. 6 *Schematic drawing of an anode installed in a hydraulic access fitting.*

one with a 6 Mo SS at two different flow rates and at varying chlorine levels at 25°C [7], and the other with the highly alloyed stainless steel 654 SMO from Avesta Shefield [9] at different temperatures, with and without pre-exposure of the specimens and with the variables of the first test.

The results from both tests are evaluated and summarised in the simplified and smoothed cathodic polarisation curves in Fig. 3 for the potential region investigated (−600 mV to +500 mV (SCE)).

The second potentiostatic test was performed within the MAST project with the stainless steel 654 SMO from Avesta Shefield [9]. Two series of specimens were used. In one the specimens were pre-exposed at the zero current potential for two days and thereafter polarised potentiostatically to different potentials between +500 and −600 mV. The other series of specimens were polarised to the same potentials from start of the test. Total exposure time was 103 days. The scatter of the current densities obtained at the chlorination level 0.4–0.6 ppm at different time intervals between 150 and 2472 h of exposure is shown in Fig. 7. The stable values are supposed to be close to the lowest values for the two series of specimens. These results for the pre-exposed specimens are in good agreement with the results from the first test [7]. For the specimens polarised from the start, however, considerably higher current densities were recorded. The results indicate that the potential/time history prior to polarisation has a great influence on the current densities recorded for potentials above −200 mV (SCE), and that the stable current densities in the potential range −400 to −600 mV (SCE) were considerably lower than previously estimated.

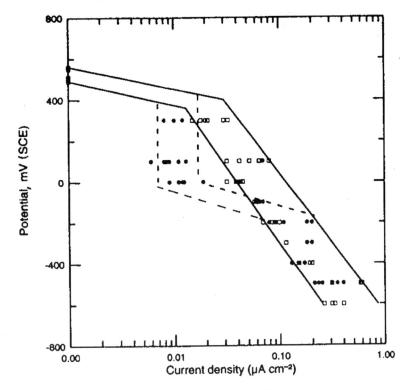

Fig. 7 Scatter of polarisation data for exposure times of 150–2472 h for a chlorination level of 0.4–0.6 ppm at 30°C [9]. • *Specimens pre-exposed at zero current potential for 48 h.* □ *Specimens polarised from start.*

Figure 8 shows variations of current density dependent on exposure time and interruption of exposure for the potentials –200, 0, and +100 mV (SCE).

Figure 9 shows that there is no effect of the flow rate for values between 0.1 and 3.2 ms^{-1}.

The main conclusions from the tests are [9]:

- The cathodic current density on stainless steel in chlorinated sea water depends on the potential/time history of the material. The current density was considerably reduced for specimens pre-exposed at zero current potential for two days compared to specimens cathodically polarised to potentials between +300 and –100 mV (SCE) from the start of the test.

- The cathodic current density is relatively high at the start of polarisation of stainless steel in chlorinated sea water, but decreases rapidly during the first 100–200 h of exposure. Interruption of exposure increases the subsequent current requirements.

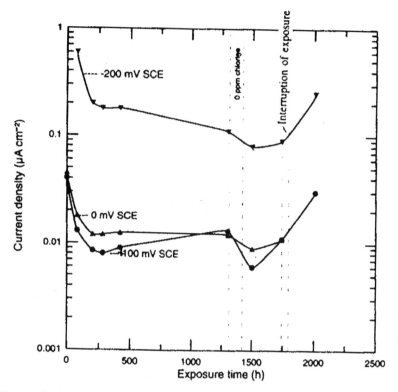

***Fig. 8** Cathodic current densities for the potentials –200 mV (SCE), 0 mV (SCE), and +100 mV (SCE) vs exposure time in periods with 30 °C and 0.4–0.5 ppm free residual chlorine for specimens not polarised the first 48 h of the test [9].*

- The cathodic polarisation current density is nearly proportional to the concentration of free residual chlorine in the potential region from +500 mV (SCE) to –600 mV (SCE) in air saturated chlorinated sea water.

- The cathodic current density in the potential region between +500 and –600 mV (SCE) was completely independent of the flow velocity in chlorinated sea water within the flow range 0.1–3.2 ms⁻¹.

- The cathodic current density on stainless steel increases with increasing temperature.

The required cathodic current density to obtain the required protection varies as shown above, and is dependent on several factors. In the design of a RCP application it is therefore important to know the service history of the system, if this is (or has been) in operation. For design of RCP it is, important, both for new and old systems, to evaluate the expected potential–time development in the stainless steel system for the anode with the resistor value chosen. The potential may, for a certain time

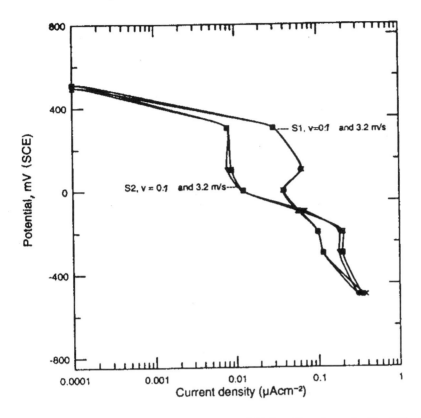

Fig. 9 Polarisation data for obtained in periods with high and low flow velocities. Chlorination: 0.5 ppm. S1 = Specimens polarised from start of the test. S2 = specimens not polarised the first 48 h [9].
• *S1: 3.2 m s⁻¹. 280 h; ▼ S2:3.2 m s⁻¹. 280 h; × S1: = 0.1 m s⁻¹. 460 h;■S2: 0.1 m s⁻¹. 460 h.*

period, exceed the safe potential if relatively high current densities are not used. This time period should be short so as not to allow crevice corrosion to initiate. If the probability for corrosion initiation is very high in this period, the need for potential reduction afterwards, below the repassivation potential to terminate the corrosion, should be evaluated. A relatively short period without chlorination in combination with RCP can contribute to the the potential reduction required.

4.1. Natural Sea Water Without Chlorination

Cathodic protection (CP) of stainless steels in natural sea water without chlorination is possible. SINTEF Corrosion and Surface Technology has designed a series of such systems to obtained protection of piping systems made of AISI 316 L SS and the systems have functioned well for many years, except for the first one in which the increased cathodic current requirements resulting from the effect of the biological slime layer was not taken into account. There is a great difference between use of RCP in chlorinated sea water and the use in natural sea water at temperatures below 30–40°C. The reason is seen in Fig. 10. The current density required to polarise stainless steel down to a safe potential of about +100 mV (SCE) is about 6 μA cm^{-2} (60 mA m^{-2}) for natural sea water contrary to the required current density in chlorinated sea water, which is less than 0.01 μA cm^{-2} (0.1 mA m^{-2}) at 0.5 ppm free residual chlorine.

The high current demand means that the possible distance for protection with one anode in a piping system will be much smaller than for an anode in a chlorinated sea water system. The anode consumption will increase in proportion to the current density, and will be orders of magnitudes higher. It is, however, found that the cathodic current density for stainless steels in natural sea water will decrease with very long exposure times, and that values after several years of exposure are much smaller than indicated by the polarisation data obtained after 14 weeks of potentiostatic polarisation, as shown in Fig. 10.

The high cathodic current densities on stainless steel in natural sea water are the result of the effect of the biological slime layer on the cathodic reaction [10–12].

The RCP method will, however, be excellent for protection of all types of stainless steel components, e.g. pumps and valves in a piping system of synthetic materials,

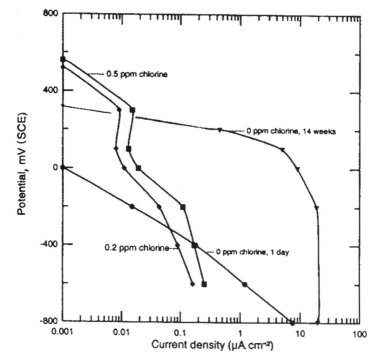

Fig. 10 Polarisation data from potentiostatic tests in chlorinated water at 30°C chlorination levels 0.2 and 0.5 ppm and exposure times of 47 and 54 days respectively [9]. Data obtained in natural sea water without chlorination at 9°C for exposure times 1 day and 14 weeks are also indicated [12].

both in natural and chlorinated sea water. Only limited areas of stainless steel will then drain current from the anodes and the anode consumption will be very small in chlorinated and moderate in natural sea water.

This method will also be excellent for the protection of stainless steels in natural sea water, for temperatures above that at which the biological slime layer has an effect. The cathodic curve will then be somewhat higher than the curve for 0 ppm chlorine after one day of exposure in Fig. 10, and the potential can then easily be suppressed to a safe potential for 6 Mo SS and other materials of similar qualities.

RCP for 6Mo SS in sea water systems with chlorination varying (for instance between the summer time and no chlorination in the winter time) may also be quite effective, even when designed with moderate current densities. In the chlorination period the potential can easily be suppressed to a safe value. In the period without chlorination the potential can be reduced, with a relatively small current density, from 300–400 mV (SCE) — which is a common potential range in natural sea water — to about 200 mV, a potential at which the risk for corrosion initiation is considerably reduced compared to the situation without RCP.

4.2. Galvanic Corrosion Prevention

The method is applicable for prevention of galvanic corrosion on less noble materials, especially for copper alloys, coupled to more noble materials like 6 Mo SS, super duplex SS and titanium alloys in chlorinated sea water. The potential on the stainless steel or titanium parts has to be polarised close to the corrosion potential of the copper alloy. Permanent polarisation to about –200 mV (SCE) should be a good compromise in terms of cathodic protection and current load on the anodes, and as indicated in Fig. 10, the required current densities are relatively low at that potential.

4.3. Local Corrosion Prevention in Produced Water Systems

The temperatures in produced water systems can be quite high, up to 50–80°C. As long as there is no oxygen in the water, common stainless steels of type AISI 316 are corrosion resistant. The potential on stainless steel in oxygen free water will be below –400 mV (SCE). However, small amounts of oxygen may increase the potential to about +200 mV (SCE), and as indicated in Fig. 1, even 6 Mo SS may corrode under these conditions. If the stainless steel AISI 316 is polarised below about –100 mV no corrosion will initiate.

The cathodic current density required to polarise stainless steel in produced water has, however, been investigated only to a small extent. No long term potentiostatic polarisation tests have been conducted. It is, however, known that the required current will be low and dependent mainly on the oxygen level and the temperature. The oxygen content in produced water will usually vary from a value of a few ppb to relative short periods in the ppm range. A CP system based on full protection at all possible oxygen levels, would probably require anodes mounted at quite short spacings. The periods with the highest oxygen levels, however, are normally rather short. A fully acceptable design would therefore require complete protection at the maximum oxygen concentration, for example 100–200 ppb. If the periods with higher oxygen content (≥ 100–200 ppb) were long enough to initiate local corrosion, then the metal would repassivate immediately when the level of oxygen was again reduced below the maximum design value.

5. Further Developments

The first RCP system was installed in a 6 Mo SS chlorinated sea water system at the Draugen platform by A/S Norske Shell [1]. The temperatures in the system is between 40 and 70°C. The system has been in service for almost one year with good results so far.

A joint R & D project is established for optimisation and for improvement of the RCP method, with respect to the design basis in chlorinated sea water and for standardisation of the anode design. The project is supported by SINTEF and CorrOcean, the oil companies Shell, Statoil and Saga, and by the Norwegian Research Council.

Statoil Sleipner has performed a laboratory loop test for 6 Mo SS in chlorinated sea water to improve the design basis and to demonstrate the possibilities of the method. A pre-study of the design, and calculation of the costs for applying RCP in the sea water systems at Sleipner A were finished in December 1995. Design of RCP for Sleipner Vest has begun and a number of smaller projects for the industry are under evaluation.

6. Conclusions

1. Field experiences have shown that high alloyed stainless steels suffer from local corrosion in sea water systems when the temperature and/or chlorination level exceed certain limits. Similar problems have been encountered for stainless steels in many other saline waters.

2. A method of cathodic protection has been developed to eliminate such corrosion problems. The method is based on resistor control of the anode current output and is called Resistor-controlled Cathodic Protection (RCP).

3. The anodes are mounted in special fittings and can be designed for lifetimes in the order of 5–30 years for chlorinated sea water systems. For a sea water system with a temperature of 30°C and a chlorination level of 0.5 ppm, the spacings between anodes in 100 and 250 mm pipes can be 28 and 44 m and for a lifetime of 10 years the anode consumptions 0.4 and 1.6 kg, respectively.

4. For existing systems the corrosion problems can be solved using the RCP method with fractions of the costs involved compared to other solutions, such as replacement of material.

5. With the RCP method the critical limits of the temperature, the chlorine level and the oxygen level can be raised considerably without any risk of corrosion.

6. The RCP system can be used in existing systems to stop corrosion, or to reduce the risk of corrosion initiation. New systems can be designed to resist more severe environmental conditions, or the system can be designed for application of less corrosion resistant materials with large material savings.

7. The most important data for reliable design of RCP systems are cathodic polarisation data for the actual materials recorded vs exposure time during long term testing in the actual environments and service conditions. For many environments, reliable data of this type of calculation for both galvanic corrosion and RCP are still lacking.

References

1. I. H. Hollen, this volume, p.128.
2. I. Nybø and S. Strandmyr, Field Experience from Statfjord and Gullfax, in *Int. Conf. on Corrosion and Materials*, Oslo, Sept 1994. Published by The Norwegian Society of Engineers (NITO).
3. J. M. Drugli and T. Rogne, Application of Stainless Steels in Offshore Industry —Design Data and Corrosion Problems. In *Welding in the World; J. Inst. Welding*, 1995, **36**, June 1995.
4. T. Rogne, J. M. Drugli and E. Bardal, Crevice Corrosion of Stainless Steel — Initiation, Propagation and Passivation Properties, in *Proc. 9th Scand. Corrosion Congr.*, Copenhagen, 1983. Published by Korrosioncentralen ATV Glostrup, Denmark.
5. Okayama *et al.*, The effect of alloying elements on repassivation potential for crevice corrosion in 3% NaCl solution, *Corros. Engng*, 1985, **36**, 157.
6. Research work performed at SINTEF, not published.
7. P. O. Gartland and J. M. Drugli, "Methods for Evaluation and Prevention of Local and Galvanic Corrosion in Chlorinated Sea water Pipelines". *Corrosion '92*, Paper No. 408, NACE, Houston, Tx, 1992.
8. P. O. Gartland, R. Johnsen and J. M. Drugli, "The RCP Method for prevention of Local Corrosion on Stainless steels in Saline Water Piping Systems", *Int. NACE Conf. On Corrosion in Natural and Industrial Environments. Problems and Solutions.* Grado, Italy, May 1995.
9. J. M. Drugli and T. Rogne, Cathodic Properties of Stainless Steel in Chlorinated Sea Water, SINTEF Report No. STF F95378, 1995-11-24.
10. A. Molica, A. Trevis, E. Traverso, G. Ventura, G. De Carolis and R. Dellepiane, Cathodic performance of stainless steels in natural sea water as a function of microorganism settlement and temperature, *Corrosion*, 1989, **45**, (1).
11. E. Bardal, J. M. Drugli and P. O. Gartland, "The Behaviour of Corrosion Resistant Materials in Sea water", *Corros. Sci.*, 1993, **35**, 1–4, 257–267.
12. R. Holthe, E. Bardal and P. O. Gartland, "Time Dependence of Cathodic Properties of Materials in Sea water", *Mat. Perform.*, 1989, **28** (6), 16.

Index